Cambridge International

AS & A Level Mathematics:

Probability
& Statistics 1

Worked Solutions Manual

CAMBRIDGE
UNIVERSITY PRESS

CAMBRIDGE
UNIVERSITY PRESS

University Printing House, Cambridge CB2 8BS, United Kingdom

One Liberty Plaza, 20th Floor, New York, NY 10006, USA

477 Williamstown Road, Port Melbourne, VIC 3207, Australia

314–321, 3rd Floor, Plot 3, Splendor Forum, Jasola District Centre, New Delhi – 110025, India

103 Penang Road, #05-06/07, Visioncrest Commercial, Singapore 238467

Cambridge University Press is part of the University of Cambridge.

It furthers the University's mission by disseminating knowledge in the pursuit of education, learning and research at the highest international levels of excellence.

www.cambridge.org
Information on this title: www.cambridge.org/9781108613095

© Cambridge University Press 2019

This publication is in copyright. Subject to statutory exception and to the provisions of relevant collective licensing agreements, no reproduction of any part may take place without the written permission of Cambridge University Press.

First published 2019

20 19 18 17 16 15 14 13 12 11 10 9 8 7 6 5 4

Printed in Great Britain by CPI Group (UK) Ltd, Croydon CR0 4YY

A catalogue record for this publication is available from the British Library

ISBN 9781108613095 Paperback with Digital Access

Cambridge University Press has no responsibility for the persistence or accuracy of URLs for external or third-party internet websites referred to in this publication, and does not guarantee that any content on such websites is, or will remain, accurate or appropriate. Information regarding prices, travel timetables, and other factual information given in this work is correct at the time of first printing but Cambridge University Press does not guarantee the accuracy of such information thereafter.

All worked solutions within this resource have been written by the author.
In examinations, the way marks are awarded may be different.

NOTICE TO TEACHERS IN THE UK

It is illegal to reproduce any part of this work in material form (including photocopying and electronic storage) except under the following circumstances:
(i) where you are abiding by a licence granted to your school or institution by the Copyright Licensing Agency;
(ii) where no such licence exists, or where you wish to exceed the terms of a licence, and you have gained the written permission of Cambridge University Press;
(iii) where you are allowed to reproduce without permission under the provisions of Chapter 3 of the Copyright, Designs and Patents Act 1988, which covers, for example, the reproduction of short passages within certain types of educational anthology and reproduction for the purposes of setting examination questions.

Contents

The items in orange are available on the Elevate edition that accompanies this book.

How to use this book	**v**
1 Representation of data	**1**
Exercise 1A	1
Exercise 1B	2
Exercise 1C	6
Exercise 1D	10
End-of-chapter review exercise 1	
2 Measures of central tendency	**12**
Exercise 2A	12
Exercise 2B	13
Exercise 2C	15
Exercise 2D	16
Exercise 2E	18
End-of-chapter review exercise 2	
3 Measures of variation	**23**
Exercise 3A	23
Exercise 3B	25
Exercise 3C	28
Exercise 3D	30
Exercise 3E	32
End-of-chapter review exercise 3	
Cross-topic review exercise 1	
4 Probability	**33**
Exercise 4A	33
Exercise 4B	34
Exercise 4C	35
Exercise 4D	38

Exercise 4E	40
Exercise 4F	41
End-of-chapter review exercise 4	

5 Permutations and combinations 45

Exercise 5A	45
Exercise 5B	45
Exercise 5C	46
Exercise 5D	47
Exercise 5E	49
Exercise 5F	50
Exercise 5G	52
End-of-chapter review exercise 5	

Cross-topic review exercise 2

6 Probability distributions 55

Exercise 6A	55
Exercise 6B	58
End-of-chapter review exercise 6	

7 The binomial and geometric distributions 63

Exercise 7A	63
Exercise 7B	66
Exercise 7C	68
Exercise 7D	69
End-of-chapter review exercise 7	

8 The normal distribution 71

Exercise 8A	71
Exercise 8B	72
Exercise 8C	73
Exercise 8D	77
Exercise 8E	80
End-of-chapter review exercise 8	

Cross-topic review exercise 3

Practice exam-style paper

How to use this book

This resource contains worked solutions to the questions in the *Cambridge International AS & A Level Mathematics: Probability & Statistics 1 Coursebook*. Both the book and accompanying Elevate edition include the solutions to the chapter exercises. You will find the solutions to the end-of-chapter review exercises, cross-topic review exercises and practice exam-style paper on the Elevate edition only.

Each solution shows you step-by-step how to solve the question. You will be aware that often questions can be solved by multiple different methods. In this book, we provide a single method for each solution. Do not be disheartened if the working in a solution does not match your own working; you may not be wrong but simply using a different method. It is good practice to challenge yourself to think about the methods you are using and whether there may be alternative methods.

All worked solutions in this resource have been written by the author. In examinations, the way marks are awarded may be different.

Additional guidance is included in **Commentary** boxes throughout the book. These boxes often clarify common misconceptions or areas of difficulty.

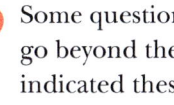

E Some questions in the coursebook go beyond the syllabus. We have indicated these solutions with a red line to the left of the text.

Chapter 1
Representation of data

EXERCISE 1A

1

0	1 2 3 3 4 4 5 6 7 8 9
1	0 1 2 3 3 5 6
2	0 6

Key: **1** | 0 represents 10 visits

Values from 1 to 26 can be shown ascending left-to-right in three rows (with digits aligned in columns) for classes of width 10, namely 0–9, 10–19 and 20–29. A suitable key must be included.

2 a

15	0 2 6 8 9
16	0 2 3 5
17	0 2 5

Key: **15** | 0 represents 150 coins

b The greatest possible value will be when as many bags as possible contain as many $1 coins as possible (nine bags), and one bag contains each of the lower value coins with the number of coins increasing as the coin value increases.

Coin value ($)	0.10	0.25	0.50	1.00
No. bags	1	1	1	9
No. coins	150	152	156	1484
Value ($)	15	38	78	1484

Greatest possible value is
$15 + 38 + 78 + 1484 = \$1615$.

3 a The most common number of employees is 18.

b Eight companies have fewer than 25 employees.

c Four of the 20 companies have more than 30 employees: $\frac{4}{20} \times 100 = 20\%$.

d i 30–39 (third row)

ii First row:
$0 + (4 \times 8) + (2 \times 9) + (7 \times 10) = 120$

Second row:
$0 + 5 + (2 \times 6) + (2 \times 7) + 8 + 9 + (8 \times 20)$
$= 208$

Third row:
$0 + (2 \times 1) + 2 + 9 + (5 \times 30) = 163$

The first row (10–19) contains the fewest employees.

4 a $649 - 561 = 88$

b $(561 \times 12.5) - 649x = 3.30$ gives $x = \$10.80$

c Least is 0 and greatest is 3.

There are three common numbers of passengers (30, 43 and 45) and these may or may not have been carried on the same days.

5 a

Batsman P		Batsman Q
	2	0 1
9 8 7 7 6	3	1 6
8 7 4 1 1	4	2 5 8
9 9 7 3 2	5	1 2 6 7
	6	4 8
	7	1 7

Key: 6 | **3** | 1 represents 36 runs for P and 31 runs for Q

Values from 20 to 77 can be shown in six rows for classes of width 10, namely 20–29, 30–39, 40–49, 50–59, 60–69 and 70–79. Values to the right of the stem ascend left-to-right and values to the left of the stem ascend right-to-left. A suitable key must be included.

b i Totals are 739 for Q and 688 for P, so Q performed better.

ii P's scores are spread from 36 to 59; Q's scores are spread from 20 to 77.
P performed more consistently because his scores are less spread out than Q's.

6 a

Wrens (10)		Dunnocks (10)
3	1	
9 8 7	1	7 9
4 3 3 2 1 0	2	2 2 3 4
	2	5 7 8
	3	0

Key: 8 | 1 | 9 represents 18 eggs for a wren and 19 eggs for a dunnock

Some numbers appear in the stem twice because the row widths are 5 rather than 10.

b $0.92 \times 237 = 218$

c $\dfrac{200 - 14}{200} \times 100 = 93\%$

7 | 81 82 84 85 | 86 86 88 89 90 90 91 91 | 92 | 93 93 | 93 94 94 95 96 | 96 97 97 98 99 |

Key: The girls' and boys' scores appear in red and blue, respectively.

The student in the middle is the 13th one from either end. That is the girl who scored 92%.

If the girl who scored 93 is placed to the left of the two boys who scored 93, as above, this results in a block of seven boys with no girls between them. So, the greatest possible number of boys not next to a girl is five.

EXERCISE 1B

1 a Lower boundary is $200 - \dfrac{1}{2} \times 50 = 175$ years;

upper boundary is $300 + \dfrac{1}{2} \times 50 = 325$ years.

b $175 - 25 = 325 - 175 = 475 - 325$
$= 625 - 475 = 150$ years

This is the width of all the intervals.

c

Frequency densities such as 15, 18, 12 and 6 buildings per 150 years or 5, 6, 4 and 2 buildings per 50 years could be used instead.

d Area of the part-columns between 250 and 400 years is:
$(325 - 250) \times 0.12 + (400 - 325) \times 0.08$
$= 9 + 6 = 15$ buildings

The units for area are $\dfrac{\text{buildings}}{1 \text{ year}} \times \text{years} = \text{buildings}$.

2 a $p = 690 - (224 + 396) = 70$

b

Frequency densities such as 56, 66 and 35 samples per 2 grams could be used instead.

c $\frac{1}{2} \times 224 + \frac{1}{2} \times 396 = 310$ samples or

$(12 - 8) \times 28 + (18 - 12) \times 33 = 310$ samples

> 8 and 18 are the mid-values of the first and second classes, whose respective frequencies are 224 and 396.

3 a $11 + 22 = 33$

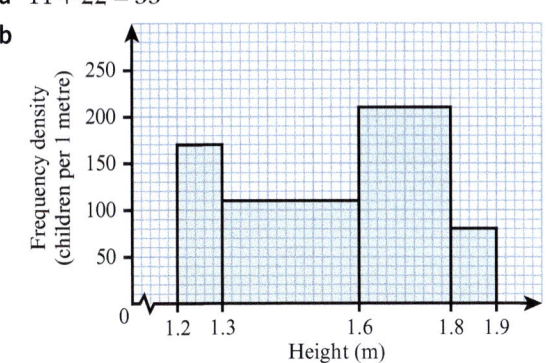

   b

   [Histogram: Frequency density (children per 1 metre) vs Height (m). Bars at 1.2–1.3 (≈170), 1.3–1.6 (≈110), 1.6–1.8 (≈210), 1.8–1.9 (≈80). Y-axis marked 50, 100, 150, 200, 250.]

   c Boys $= \frac{1}{2} \times 26 + 6 = 19$. Girls $= \frac{1}{2} \times 16 + 2 = 10$. Total is $19 + 10 = 29$ children.

4 a Any u from 35 to 50.

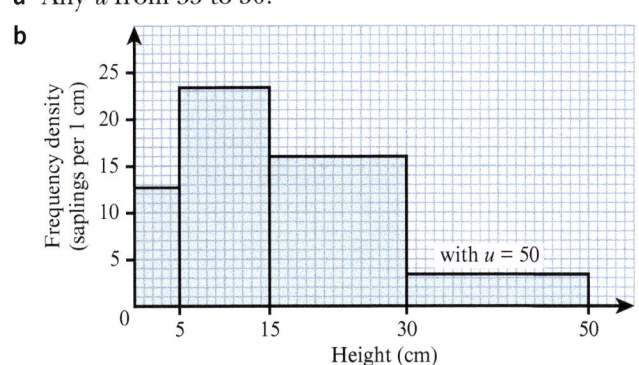

   b

   [Histogram with $u = 50$: Frequency density (saplings per 1 cm) vs Height (cm). Bars: 5–15 (≈13 then ≈23), 15–30 (≈16), 30–50 (≈4). Y-axis marked 5, 10, 15, 20, 25.]

   c i All of the first two classes plus two-thirds of the third class.

   $64 + 232 + \left(\frac{25 - 15}{30 - 15} \times 240\right) = 296 + \left(\frac{2}{3} \times 240\right) = 456$

   ii Three-quarters of the second class plus nine-thirtieths of the third class.

   $\left(\frac{15 - 7.5}{15 - 5} \times 232\right) + \left(\frac{19.5 - 15}{30 - 15} \times 240\right) = \left(\frac{3}{4} \times 232\right) + \left(\frac{9}{30} \times 240\right) = 246$

Cambridge International AS & A Level Mathematics: Worked Solutions Manual

5 **a** Class boundaries are 2.55 and 2.85 minutes, and 2.85 − 2.55 = 0.3 minutes.

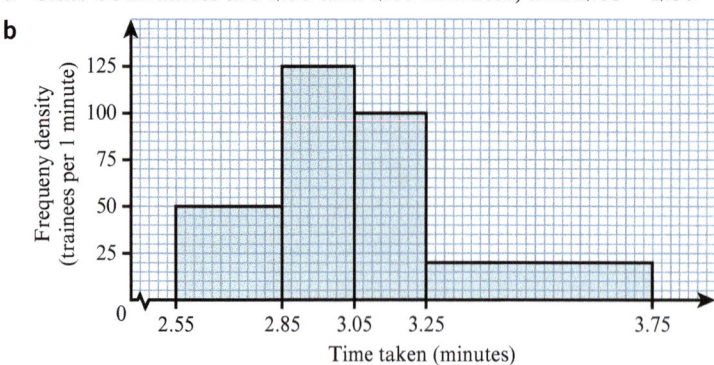

b

[Histogram with x-axis "Time taken (minutes)" showing boundaries at 2.55, 2.85, 3.05, 3.25, 3.75 and y-axis "Frequency density (trainees per 1 minute)" with values 0, 25, 50, 75, 100, 125. Bars: 2.55–2.85 at 50, 2.85–3.05 at 125, 3.05–3.25 at 100, 3.25–3.75 at about 20.]

c $2.55 + \left[\frac{10}{15} \times (2.85 - 2.55)\right] = 2.55 + \left[\frac{2}{3} \times 0.3\right] = 2.75$ minutes or 2 minutes 45 seconds or 165 seconds

> The fastest ten trainees make up two-thirds of the first class. Two-thirds of the way along the interval from 2.55 to 2.85 is at 2.75.

d i $b = 3.25 + \left[\frac{5}{10} \times (3.75 - 3.25)\right] = 3.25 + \left[\frac{1}{2} \times 0.5\right] = 3.5$

> Ten of the 15 trainees are in the third class, so five of the 15 are in the fourth class, which is half of the trainees in that class. Half-way along the interval from 3.25 to 3.75 is at 3.5.

  ii $b = 3.05 - \left[\frac{5}{25} \times (3.05 - 2.85)\right] = 3.05 - \left[\frac{1}{5} \times 0.2\right] = 3.01$

6 **a** $(2 \times 12) + (2 \times 18) + (8 \times 16) + (6 \times 20) + (2 \times 8) = 324$

> Total area of the columns is equal to the total frequency of the data.

  b i $\left(\frac{1}{2} \times 24\right) + \left(\frac{1}{2} \times 36\right) = 12 + 18 = 30$

  ii $\left(\frac{2}{8} \times 128\right) + \left(\frac{3}{6} \times 120\right) = 32 + 60 = 92$

  c Let the total number be n, then $0.15n = 324$, which gives $n = 2160$.

  d Monitored and delayed by 3 to 7 minutes $\approx \left(\frac{1}{2} \times 36\right) + \left(\frac{3}{8} \times 128\right) = 18 + 48 = 66$

  Let the number of August journeys delayed by 3 to 7 minutes be x, then $0.15x = 66$, so $x = 440$.
  We assume that the proportion of all August journeys delayed by 3 to 7 minutes is the same as the proportion of monitored journeys delayed by 3 to 7 minutes.

7 **a** $(20 \times 2) + (40 \times 3) + (60 \times 4) + (80 \times 1) = 480$

  b $40 + \frac{50 - 20}{60 - 20} \times 120 = 130$

Chapter 1: Representation of data

  c 25% of 480 = 120 hard drives
  $\left(\dfrac{120-k}{120-60} \times 240\right) + 80 = 120$, so $k = 120 - \dfrac{(120-80)(120-60)}{240} = 110$

8 a $\dfrac{10 \times 5 + 15 \times 25}{575} = \dfrac{17}{23}$

  b $\left(\dfrac{25-12.4}{25-10} \times 375\right) + 50 + \left(\dfrac{36.8-30}{50-30} \times 100\right) = 399$

  c 20% of 575 = 115, so the 116th item is the shortest that will not be recycled.
  Estimate is $10 + \left[\dfrac{116-50}{375} \times (25-10)\right] = 12.6$ cm

> There are two ways to calculate the estimate:
> Lower class boundary + appropriate fraction of the interval width or upper class boundary − appropriate fraction of the interval width.
> The two 'appropriate fractions' sum to 1.

9 a $(0.3 \times 4) : (0.4 \times 2) : (0.3 \times 1) = 1.2 : 0.8 : 0.3 = 12 : 8 : 3$

  b Area of first column is
  $(0.4 - 0.1)$ mm $\times \dfrac{4 \text{ sheets}}{n \text{ mm}} = \dfrac{1.2}{n}$ sheets
  $\dfrac{1.2}{n} = 180$ gives $n = \dfrac{1}{150}$

  c i $180 + (0.1 \times 2 \times 150) = 210$

  ii $(0.05 \times 2 \times 150) + (0.14 \times 1 \times 150) = 15 + 21 = 36$

  d Ratio thin : medium : thick is 69 : 207 : 69.
  a is in the first class and $(0.4 - a) \times 4 \times 150 = 180 - 69$ gives $a = 0.215$.
  b is in the second class and $(b - 0.4) \times 2 \times 150 = 207 - (180 - 69)$ gives $b = 0.720$.
  Medium thickness sheets are such that $0.215 \leq k < 0.720$ mm.
  We can only be certain that $0.1 \leq a < 0.4$ and that $0.4 \leq b < 0.8$.

10 a Frequency density is $\dfrac{371}{7} = \dfrac{1060}{20} = 53$ for all classes, so class frequency = 53 × class width.
  $a = 53 \times 3 = 159$ and $b = 53 \times 12 = 636$
  Total frequency is $159 + 371 + 1060 + 636 = 2226$.

  b 50% of 2226 = 1113, so the 1114th mass is the lightest of the heaviest 50%.
  $1114 - 159 - 371 = 584$th mass in third class
  Estimate is $12.5 + \left[\dfrac{584}{1060} \times (32.5 - 12.5)\right] = 23.5$ kg.

11

Class	Width	f	fd	Height
0.5 to 2	2.25 − 0.25 = 2	n	$\dfrac{n}{2}$	h
−2.5 to −0.5	−0.25 − −2.75 = 2.5	d	$\dfrac{d}{2.5}$	x

> Column heights are proportional to (in the same ratio as) frequency densities.

The ratios $x : h$ and $\dfrac{d}{2.5} : \dfrac{n}{2}$ are equal, so $\dfrac{x}{h} = \dfrac{\frac{d}{2.5}}{\frac{n}{2}}$, which gives $x = \dfrac{4hd}{5n}$.

12 Ratio of column widths is $\dfrac{10}{4} : \dfrac{15}{3} : \dfrac{24}{2} : \dfrac{8}{1} = 3 : 6 : 14.4 : 9.6$.

Total width is $3 + 6 + 14.4 + 9.6 = 33$ cm.

> Multiply the ratio of column widths by 1.2, so that the smallest number becomes 3, which is the width of the narrowest column.

13

Frequency	165	240	195	147
Width	$50 - p + 1$ $= 51 - p$	20	10	$q - 81 + 1$ $= q - 80$
fd	$\dfrac{165}{51 - p}$	12	19.5	$\dfrac{147}{q - 80}$

> $5 : 8 : 13 : 7$ is equivalent to $7.5 : 12 : 19.5 : 10.5$, and these are the frequency densities.

$\dfrac{165}{51 - p} = 7.5$ gives $p = 29$ and $\dfrac{147}{q - 80} = 10.5$ gives $q = 94$.

EXERCISE 1C

1 a

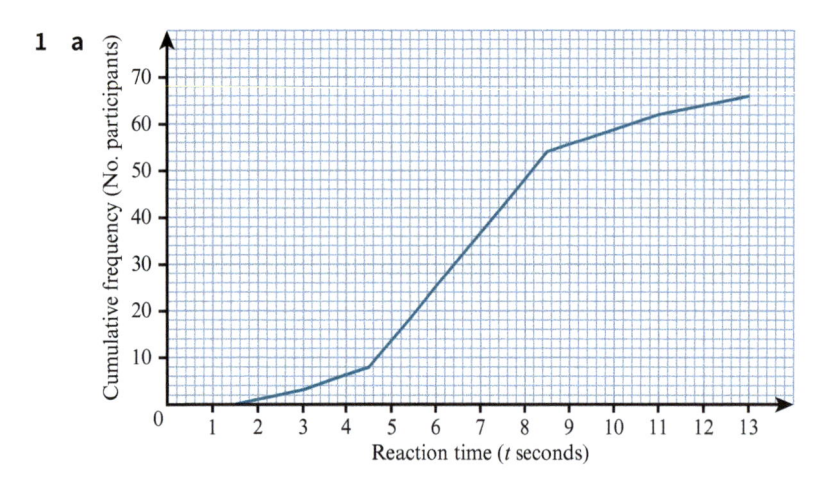

b i 7.5 seconds is the 43rd value and 5.5 seconds is the 20th value: $43 - 20 = 23$.

ii The slowest 20 times are for participants who took longer than the $66 - 20 = 46$th, and the 46th value is 7.8 seconds.

Chapter 1: Representation of data

2 a $19 + 0.5 = 19.5$ cm

b

Width (cm)	< 9.5	< 14.5	< 19.5	< 29.5	< 39.5	< 44.5
No. books (cf)	0	3	16	41	65	70

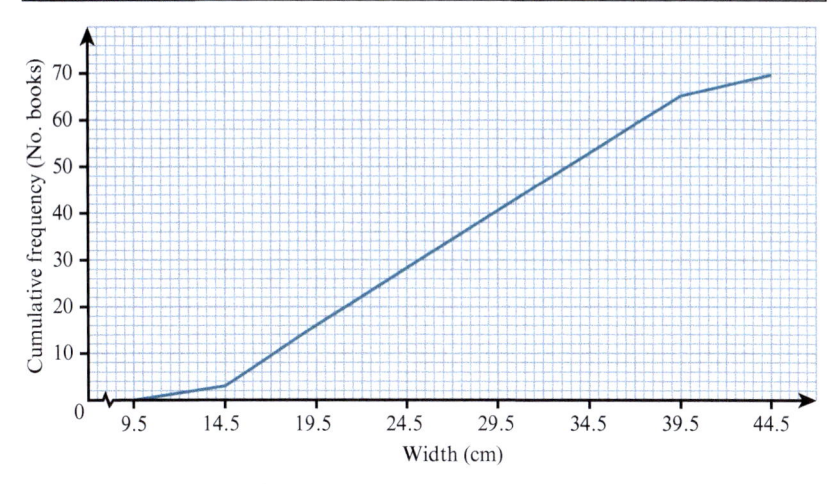

c i cf-value at width 27 cm is 34 or 35.

ii The widest 20 books are wider than the 50th book. The 50th value is 33.25 cm and the 70th (last) value is 44.5 cm, so the widths are from ≈ 33.25 to 44.5 cm.

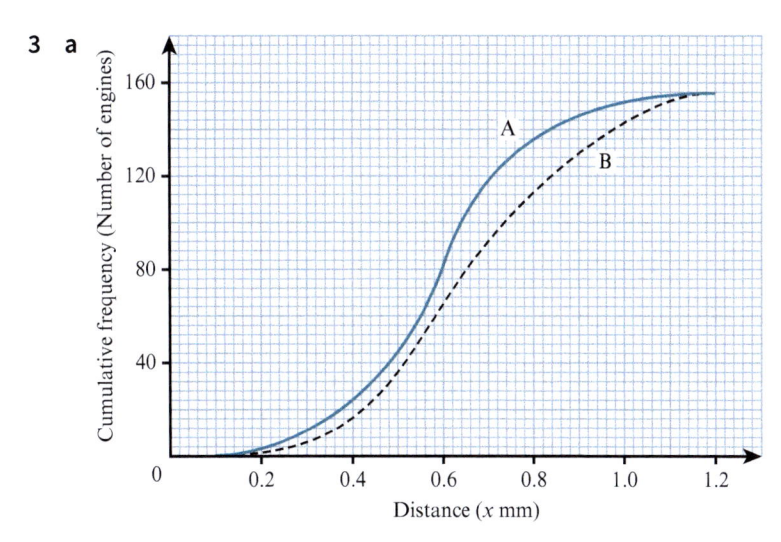

3 a

b i ≈ 118 − 11 = 107 for type A engines and ≈ 91 − 4 = 87 for type B engines.

ii ≈ 62 type A engines + 46 type B engines = 108 engines

c 16 type A engines have a measurement of more than 0.81 mm, so this is the fixed amount. The number of type B engines with measurements of more than 0.81 mm is ≈ 42.

4 a $37 - 20 = 17$

Points are plotted at $(0.2, 20)$ and $(0.4, 37)$, so we know that these cf-values are precise.

b i 12

ii 32 radii are 0.16 cm or less ($d \leq 0.32$ cm), so $60 - 32 = 28$ have $r > 0.16$ cm.

c 80% of $60 = 48$ components have diameters less than k, so $k = 4.7$ to 4.8.

d It has the highest frequency density, and is where the graph is steepest.

5 a i $132 - 68 = 64$

ii $124 - 48 = 76$

b $\approx 14.0 - 6.6 = 7.4$ g

c $152 \times 2 = 304$ cut diamonds all have masses less than $24 \div 2 = 12$ grams. Point is at $(12, 304)$.

6 $a = 32$; $b = 77 - 32 = 45$; $c = 92 - 77 = 15$ and $d = 125 - 92 = 33$.

7 a $76 - 11 = 65$

b $\dfrac{x+y}{2} = \dfrac{30 + 20}{2} = 25$ minutes

Number of staff that takes more than 25 minutes is $80 - 56 = 24$.

$x = 30$ (68th value) and $y = 20$ ($80 - 56 = 24$th value).

8 a The heights are not representative of 12-year-olds in general *or* The ratio of under-155 cm to over-155 cm is 3 : 1 for boys and 1 : 3 for girls.

b $100 - (18 \text{ or } 19) = 81$ or 82

c There are equal numbers of boys and girls below and above this height.

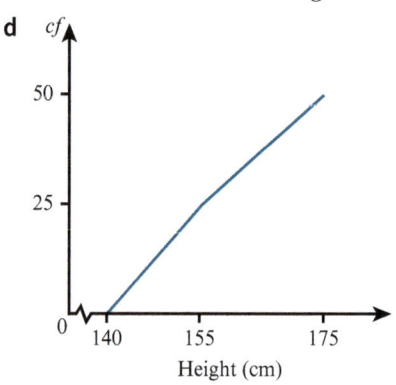

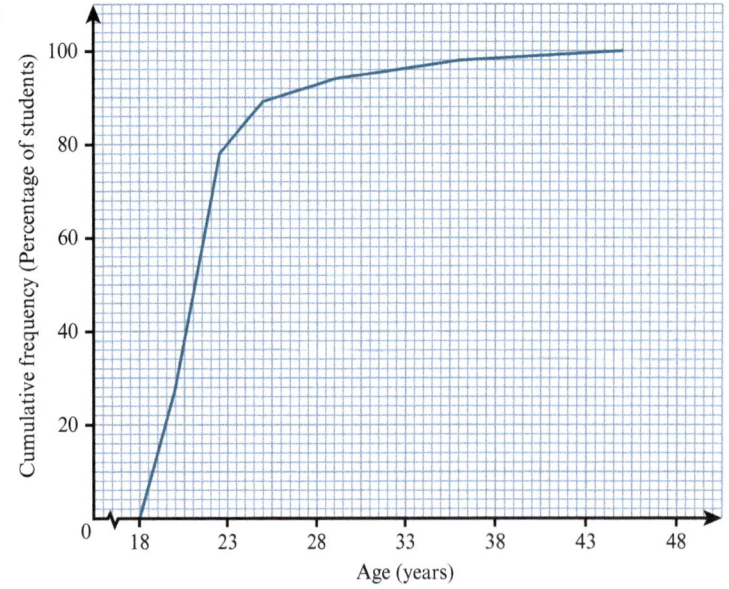

d

9 a

b Age at 92% is 27.3 to 27.4 years.

Mature students are those over the age of 27 years and 4 or 5 months.

c i $\dfrac{324-54}{0.27} = 1000$

  ii We assume that all age groups are equally likely to find employment.

  It could be an under-estimate because, for example, older graduates may have some work experience which makes them more attractive to employers.

  It could be an over-estimate because, for example, the economic situation deteriorates before the students complete their courses.

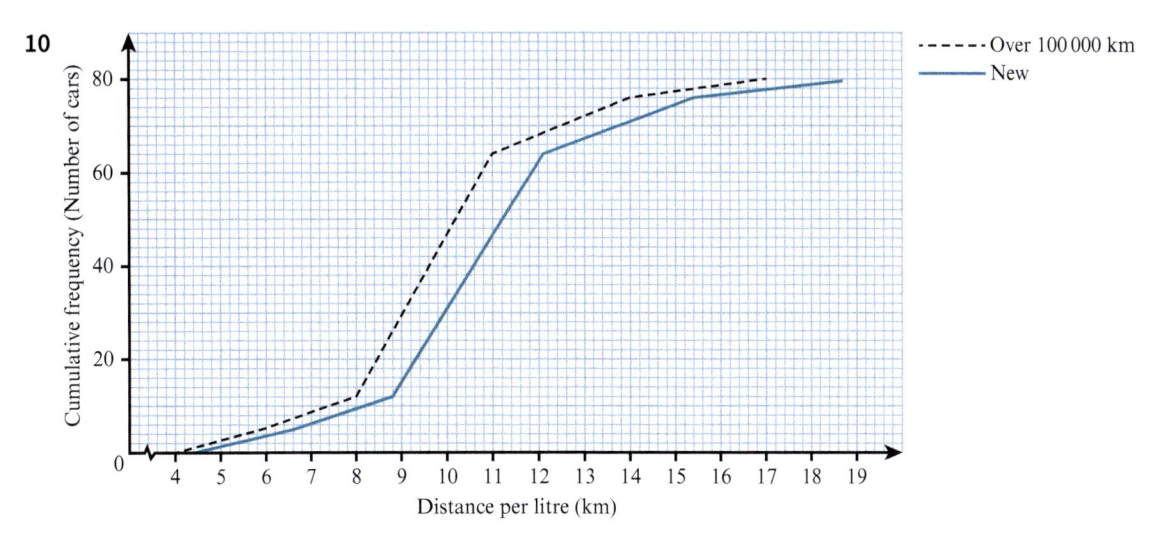

10

Polygons give approximately 17 cars and curves give approximately 16 cars.

> Find the difference between *cf*-values at 10.5 km on the graphs.

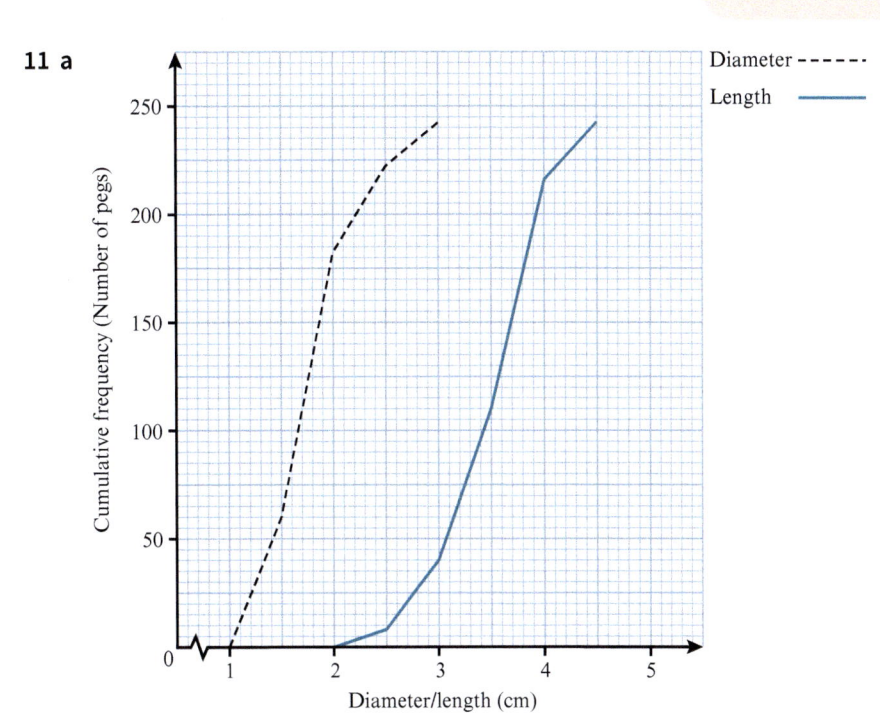

11 a

b Least possible n is 0 and the greatest possible n is $8 + 20 = 28$.

c Diameter and length for individual pegs are not shown.

	Acceptable	Unacceptable
Length	215	27
Diameter	198	44

> At most eight pegs have lengths and diameters from 2.0 to 2.5 cm.
> At most 20 pegs have lengths and diameters from 2.5 to 3.0 cm.

Least possible number acceptable is $215 - 44 = 171$.

Greatest possible number acceptable is 198.

Best estimate is 'between 171 and 198 inclusive'.

The length and diameter of each peg should be recorded together, then the company can decide whether each peg is acceptable or not.

EXERCISE 1D

1 a Any suitable for qualitative data (pictogram, pie chart, bar chart, etc.)

b Pie chart, as $\frac{3}{4}$ of a circle is easily recognised, or a sectional percentage bar chart.

2 Histogram; area of middle three columns is greater than half total column area.

3 a The numbers can be shown in compact form on three rows, but a bar chart requires 17 bars, all with frequencies of 0 or 1.

b Seller may notice that $12 + 16 + 7 + 8 + 21 + 4 + 11 + 6 + 5 + 10 = 100$, which means that 11 (rather than ten) boxes of 100 tiles could be offered for sale.

4 a

Month	1st	2nd	3rd	4th	5th	6th	7th	8th
% raised	36	32	16	8	4	2	1	0.5
Cumulative % raised	36	68	84	92	96	98	99	99.5

It will take seven months.

b Percentage cumulative frequency graph; the graph passes below the point (12, 100).

5 a Frequency density may be mistaken for frequency in the histogram, so it may be seen to represent a total of $12 + 9 + 8 = 29$ trees.

Pic chart does not show the actual numbers of trees.

b Pictogram: one symbol to represent six trees.

Short (two symbols); medium (three symbols); tall (four symbols), plus a key.

It will show the actual numbers of trees.

6 a

Table 1							
Score (%)	30–39	40–49	50–59	60–69	70–79	80–89	90–99
Frequency	3	5	6	15	5	4	2

b

Table 2			
Grade	C	B	A
Frequency	8	26	6

Any three valid, non-zero frequencies that sum to 40, such as three Cs, 35 Bs and two As.

c Raw: stem-and-leaf diagram is appropriate.

Tables 1 and 2 do not show raw marks, so stem-and-leaf diagrams are not appropriate.

Table 1: any suitable for grouped discrete data, e.g. a histogram.

Table 2: any suitable for qualitative data.

7 a For example, he worked for less than 34 hours in 49 weeks, and for more than 34 hours in three weeks.

b It may appear that Tom worked for more than 34 hours in a significant number of weeks.

c Histogram: boundaries at 9, 34 and 44 with frequency densities of 98 and 15.

Pie chart: sector angles ≈ 339.2° and 20.8°.

Bar chart: frequencies 49 and 3.

Sectional percentage bar chart: ≈ 94.2% and 5.8%.

8 a Some classes overlap *or* Upper boundary of one class ≠ lower boundary of the following class or classes are not continuous.

b Refer to the five focal lengths as, say, types A, B, C, D and E in a key.

Pie chart with sector angles 77.1°, 128.6°, 77.1°, 51.4° and 25.7°.

Bar chart or vertical line graph with heights 18, 30, 18, 12 and 6.

Pictogram with a symbol representing 1, 3 or 6 lenses.

9

Country	C	SL	Ma	G	Mo
Number (hundred thousands)	24.8	18.4	10.8	7.91	7.50
% of population	14.4	8.9	3.8	17.7	27.4

Ranked in ascending order by number: Mo, G, Ma, SL, C.

Ranked in ascending order by percentage: Ma, SL, C, G, Mo.

You are expected to notice that, for example, the country with the lowest number of people living in poverty (Mongolia) has the highest proportion of people living in poverty.

Two bar charts on the same diagram, as below, show this clearly.

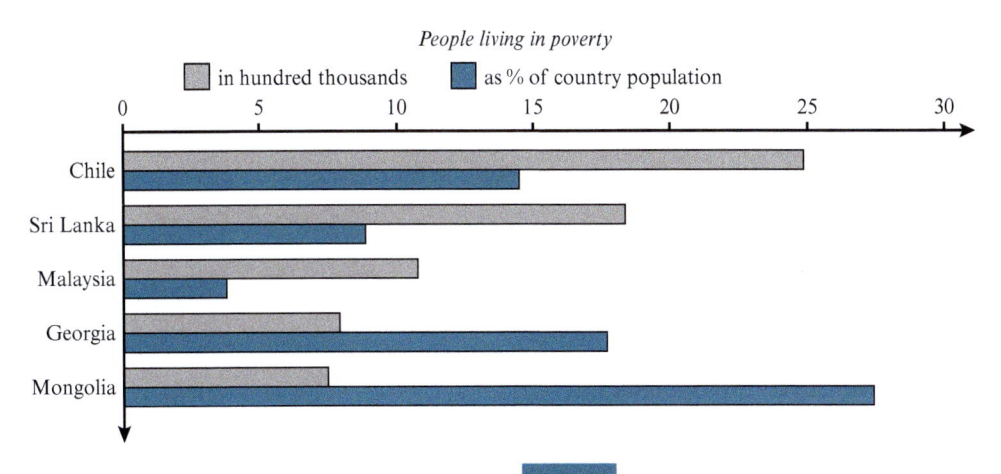

People living in poverty

Chapter 2
Measures of central tendency

EXERCISE 2A

1 a There is no mode because each value appears once.

  b The modes are 16, 19 and 21, which appear twice each.

2 The mode is 'the'. It is the only word that appears more than once.

3 The mode for x is 7 and the mode for y is -2.

> The mode is the value with the highest frequency.

4 Frequency densities for x are $\dfrac{5}{4-0}$, $\dfrac{9}{14-4}$ and $\dfrac{8}{20-14}$, i.e. 1.25, 0.9 and 1.33, so the modal class is 14–20.

Frequency densities for y are $\dfrac{66}{6.5-2.5}$, $\dfrac{80}{11.5-6.5}$ and $\dfrac{134}{20.5-11.5}$, i.e. 16.5, 16 and 14.9, so the modal class is 3–6.

> The modal class is the class with the highest frequency density.

5 The most popular size(s) can be pre-cut to serve customers more quickly. This may also result in less wastage of materials.

6 Each class has a frequency density of $\dfrac{84}{7.5-0.5} = 12$.

Sum of class frequencies = $12 \times$ sum of interval widths = $12 \times (25.5 - 7.5) = 216$.

> If there is no modal class then all classes have the same frequency density.

7 Highest of the three frequency densities is $\dfrac{27}{19.8-17.4} = 11.25$.

$\dfrac{a}{1.8} < 11.25$, so $a < 20.25$ and the largest possible a is 20.

$\dfrac{b}{2} < 11.25$, so $b < 22.5$ and the largest possible b is 22.

Largest possible n is $20 + 22 + 27 = 69$.

8 Area ratio of first and third classes is $(4 \times 4) : (8 \times 3) = 2 : 3$.

First class: frequency = $\dfrac{2}{2+3} \times 120 = 48$;

frequency density = $\dfrac{48}{4} = 12$.

Third class: frequency = $\dfrac{3}{2+3} \times 120 = 72$;

frequency density = $\dfrac{72}{8} = 9$.

Second class: frequency density > 12, so frequency > $6 \times 12 = 72$.

The least possible frequency of the modal class is 73.

Chapter 2: Measures of central tendency

EXERCISE 2B

1. a $\dfrac{400}{8} = 50$ b $\dfrac{42.6}{6} = 7.1$ c $21\dfrac{5}{8} \div 5 = 4\dfrac{13}{40}$ or 4.325

2. a $\dfrac{p^2 + 392}{7} = 63$, which gives $p = \pm 7$.

   b $\dfrac{q^2 + q + 70}{8} = 20$

   $q^2 + q - 90 = 0$

   $(q-9)(q+10) = 0$, so $q = 9$ or $q = -10$.

3. a $\bar{x} = \dfrac{\Sigma x}{n} = \dfrac{325.5}{14} = 23.25$ b $\Sigma y = n \times \bar{y} = 45 \times 23.6 = 1062$ c $n = \dfrac{\Sigma z}{\bar{z}} = \dfrac{4598}{52.25} = 88$

   d $\Sigma f = \dfrac{\Sigma xf}{\bar{x}} = 86 \div 7\dfrac{1}{6} = 12$ e $\Sigma xf = \bar{x} \times \Sigma f = 0.842 \times 135 = 113.67$

4. a $\bar{x} = \dfrac{\Sigma xf}{\Sigma f} = \dfrac{144 + 185 + 323 + 468 + 20}{60} = \dfrac{1140}{60} = 19$

   b $\bar{y} = \dfrac{\Sigma yf}{\Sigma f} = \dfrac{459.74 + 762.85 + 1184.96 + 1079.61 + 938.74}{1200} = \dfrac{4425.9}{1200} = 3.68825$

5. $\dfrac{\Sigma qf}{\Sigma f} = \dfrac{63 + 104 + 9a + 110}{a + 33} = \dfrac{77}{9}$

   $9(9a + 277) = 77(a + 33)$

   $81a + 2493 = 77a + 2541$ which gives $a = 12$.

6. a $\bar{x} = \dfrac{\Sigma xf}{\Sigma f} \approx \dfrac{1 \times 8 + 3 \times 9 + 6 \times 11 + 11 \times 2}{30} = \dfrac{123}{30} = 4.1$

   b $\bar{y} = \dfrac{\Sigma yf}{\Sigma f} \approx \dfrac{14.5 \times 7 + 18.5 \times 17 + 24.5 \times 29 + 30.5 \times 16 + 34.5 \times 11}{80} = \dfrac{1994}{80} = 24.925$

   > Use class mid-values to calculate an estimate of the mean, and remember to multiply each of them by the frequency of the class that it represents.

7. $\dfrac{71 \times 22 + 76 \times 28}{50} = 73.8\%$

8. $(1942 \times 13) - (1950 \times 12) = \$1846$

9. Age of member who left is $26\dfrac{1}{4} \times 16 - 26 \times 15 = 30$ yrs.

   This might not be very accurate because the two given means may only be accurate to the nearest month.

   > If the two given means are only accurate to the nearest month then the lower and upper boundaries of this person's possible age can be found using $26\dfrac{5}{24} \leq$ mean for $16 < 26\dfrac{7}{24}$ and $25\dfrac{23}{24} \leq$ mean for $15 < 26\dfrac{1}{24}$.

10 a The mean ($10) is not a good average because 36 of the 37 employees earn less than this.

b $\dfrac{6 \times 8 + 7 \times 11 + 8 \times 17}{36} = \dfrac{261}{36} = \$7.25$

11 a Total number of passengers is $30 \times 2 \times 61.5 = 3690$.

Class frequencies (nearest integers) are 1107, 1513 and 1070.

Total revenue is $34 \times 1107 + 38 \times 1513 + 45 \times 1070 = \$143\,282$.

This is an approximation because the frequencies 1513 and 1070 are estimates from rounded percentages. Also, we assume that 30% of 3690 = 1107 is the correct number of passengers that paid $34 each.

41% means $40.5 \leqslant \% < 41.5$, so from 1495 to 1531 passengers paid $38 each.

29% means $28.5 \leqslant \% < 29.5$, so from 1052 to 1088 passengers paid $45 each.

b Assuming that 1107 passengers paid $34 each:

Minimum (frequencies 1107, 1531, 1052) is $143 156.

Maximum (frequencies 1107, 1495, 1088) is $143 408.

Therefore $k = 143\,408 - 143\,156 = 252$.

> If we do not assume that 1107 passengers paid $34 each then:
> Minimum possible revenue (frequencies 1125, 1513, 1052) is $143 084.
> Maximum possible revenue (frequencies 1089, 1513, 1088) is $143 480.
> These give $k = 143\,480 - 143\,084 = 396$.
> As different assumptions can lead to a different value for k, it is important that we state any assumptions used as part of the solution.

12 Class frequencies are 12, 15, 21 and 6.

Short half mean (first two classes) $\approx \dfrac{142 \times 12 + 147 \times 15}{27} = 144\dfrac{7}{9}$ cm.

Tall half mean (last two classes) $\approx \dfrac{153 \times 21 + 157.5 \times 6}{27} = 154$ cm.

Difference is $154 - 144\dfrac{7}{9} = 9\dfrac{2}{9}$ or 9.22 cm.

13 a Mean $\approx \dfrac{24.5 \times 329 + 39.5 \times 413 + 64.5 \times 704 + 90 \times 258}{1704} = \dfrac{93002}{1704} = 54.6$

b Let the mean number of tomatoes per plot be t.

$1704t \times \dfrac{156.50}{1000} \times 3.2 = 50\,350$

> Note that mass is in kg throughout the equation.

$t = \dfrac{50\,350}{853.3632} = 59.0$

c The scales may be over-estimating masses *or* Revenue may not be from the sale of all the tomatoes (some may have been damaged and not arrived at market).

14 a $\dfrac{30}{20} = 1.5$

b i $\dfrac{98}{50} = 1.96$ **ii** $\dfrac{174}{50} = 3.48$

c For example, a bar chart with four groups of four bars or separate tables for boys and girls.

15 Let m be the mid-value of the class $400 - p$.

$$\frac{180 \times 12 + 260 \times 28 + 360 \times 48 + 32m}{120} = 348$$

$$26\,720 + 32m = 41\,760 \text{ gives } m = 470.$$

$$\frac{p + 400}{2} = 470, \text{ so } p = 540$$

$$\frac{(12 \times 180) + 260(28 + n) + (360 \times 48) + (470 \times 32)}{120 + n} = 340$$

$41\,760 + 260n = 40\,800 + 340n$ gives $n = 12$.

We assume that none of the 120 refrigerators have been removed from the warehouse.

---

16 Mean $\approx \dfrac{(5.5 \times 2) + (7 \times 8)}{10} = 6.7$ rooms per day,

so $\dfrac{72}{6.7} = 10.74 \ldots$ days are needed.

Therefore, one more day is required.

We assume that the work is done at the same rate and that the remaining rooms take the same amount of time, on average, as those already completed.

> Only completed rooms are counted.
> Five completed rooms means from five up to but not including six (mid-value is 5.5).
> Six or seven completed rooms means from six up to but not including eight (mid-value is 7).

17 a i $\dfrac{\sqrt{52} + \sqrt{20} + 6}{3} = 5.89$ cm

  ii $\dfrac{\sqrt{52} + \sqrt{20} + 2\sqrt{5} + 2\sqrt{2}}{3} = 5.76$ cm

b $\text{Cos}\,P\hat{O}Z = \dfrac{2^2 + 4^2 - \left(\dfrac{\sqrt{52} + \sqrt{20}}{2}\right)^2}{2 \times 2 \times 4}$

$\text{Cos}\,P\hat{O}Z = -0.882782218 \ldots$ gives $P\hat{O}Z = 152.0°$

> Use the cosine rule with $PO = 2$, $OZ = 4$ and $PZ = \dfrac{\sqrt{52} + \sqrt{20}}{2}$.

EXERCISE 2C

1 a $\Sigma x = n \times \bar{x} = 10 \times 7.4 = 74$

  b $\Sigma(x + 2) = 10 \times (7.4 + 2) = 94$

  c $\Sigma(x - 1) = 10 \times (7.4 - 1) = 64$

2 $\bar{z} = \dfrac{\Sigma(z - 7)}{25} + 7 = \dfrac{275}{25} + 7 = 11 + 7 = 18$

3 $\dfrac{\Sigma(q - 4)}{n} + 4 = 22$

$\dfrac{3672}{n} = 22 - 4$, so $n = \dfrac{3672}{18} = 204.$

4 $\bar{x} = \dfrac{\Sigma(x - 40)}{2500} + 40 = \dfrac{875}{2500} + 40 = 40.35$ mm

5 Let the sixth coded value be x, then sum of the six coded values is $28.4 + x$.

$\dfrac{28.4 + x + 6 \times 13}{6} = 17.6$

$x + 106.4 = 105.6$ gives $x = -0.8.$

6 a To show whether the cards fit into the slot ($x = w - 24 < 0$), or not ($x = w - 24 \geq 0$).

   b $\dfrac{6+2}{400} \times 100 = 2\%$

   c $\bar{w} = \bar{x} + 24 \approx \dfrac{(-0.125 \times 32) + (-0.05 \times 360) + (0.05 \times 6) + (0.15 \times 2)}{400} + 24 = \dfrac{-21.4}{400} + 24$
   $= -0.0535 + 24 = 23.9465$ mm

---

7 The maths is simpler for Fidel because his deviations are all positive, but Ramon's are all negative.

   Fidel $\bar{x} = 917 + \dfrac{15.84}{16} = 917.99$

   Ramon $\bar{x} = 920 + \dfrac{-32.16}{16} = 917.99$

8 $\Sigma t_{\text{In}} = 45 \times 60 + 83.7 = 2783.7$
   $\Sigma t_{\text{Out}} = 75 \times 65 - 38.7 = 4836.3$

   $\Sigma t = 2783.7 + 4836.3 = 7620$ and $n = 120$, so

   $\bar{t} = \dfrac{7620}{120} = 63.5$ seconds.

   This answer is accurate to one decimal place.

   Use $83.65 \leq \Sigma(t - 60) < 83.75$ and
   $-33.75 \leq \Sigma(t - 65) < -38.65$ to show
   that $63.49916 \leq$ true mean $< 63.500\dot{8}\dot{3}$.

9 a $\bar{y} = \dfrac{180n°}{3n} = 60°$

   $3n$ angles are measured and their sum is $180n°$.

   b $\Sigma(y - 30) = \Sigma y - (3n \times 30) = 180n - 90n = 90n$

10 $\bar{x} = \dfrac{58}{20} + 1 = 3.9$, so $\Sigma x = 20 \times 3.9 = 78$ and

   $\bar{y} = \dfrac{36}{30} + 2 = 3.2$, so $\Sigma y = 30 \times 3.2 = 96$.

   Mean(x and y) $= \dfrac{\Sigma x + \Sigma y}{20 + 30} = \dfrac{78 + 96}{20 + 30} = 3.48$

   Alternatively, mean(x and y)
   $= \dfrac{[58 + (20 \times 1)] + [36 + (30 \times 2)]}{20 + 30} = 3.48$.

11 $\bar{t} = \dfrac{1.44}{24} + 1.1 = 1.16$, so $\Sigma t = 24 \times 1.16 = 27.84$

   $\bar{v} = \dfrac{0.56}{16} + 1.2 = 1.235$, so $\Sigma v = 16 \times 1.235 = 19.76$

   Mean(t and v) $= \dfrac{\Sigma t + \Sigma v}{24 + 16} = \dfrac{27.84 + 19.76}{24 + 16}$

   $= \dfrac{47.60}{40} = \$1.19$

   Alternatively, mean(t and v)
   $= \dfrac{[1.44 + (24 \times 1.1)] + [0.56 + (16 \times 1.2)]}{40} = \$1.19$.

EXERCISE 2D

1 $\Sigma 1000x = 1000 \Sigma x = 1000 \times 12 \times 0.475 = 5700$; it represents the total mass in grams.

2 a 5 carats = 1 gram, so total mass is $5\Sigma x$ or $\Sigma 5x$.

   b 0.001 kg = 1 gram, so total mass is $0.001 \Sigma x$ or $\Sigma 0.001 x$.

3 1 hectare = 0.01 km², so total area is $0.01 \Sigma w$ or $\Sigma 0.01 w$.

Chapter 2: Measures of central tendency

4 $1 \text{ m/s} = 0.001 \text{ km} \div \dfrac{1}{60 \times 60} \text{ h}$

    $= 0.001 \times 60 \times 60 \text{ km/h} = 3.6 \text{ km/h}$, so $k = 3.6$

> $1 \text{ metre} = 0.001 \text{ km}$
> $1 \text{ second} = \dfrac{1}{60 \times 60} \text{ hours}$

5 **a** Calculate an estimate in mph then multiply by $\dfrac{8}{5}$ or by 1.6.

    b Mean in km/h = mean in mph × 1.6 ≈ $\dfrac{16 \times 9 + 18.5 \times 13 + 22 \times 14 + 24.5 \times 4}{40} \times 1.6 = 31.62$

6 $3\Sigma x - 30 = 528$ gives $\Sigma x = \dfrac{528 + 30}{3} = 186$, so $\bar{x} = \dfrac{186}{15} = 12.4$

    $0.5\Sigma x - 15b = 138$
    $0.5 \times 186 - 15b = 138$ gives $b = -3$

7 $125a - 20b = 400$ .... [1]
    $-20a + 125b = 545$ .... [2]   [1] × 4 + [2] × 25 gives $b = 5$ and $a = 4$

8 **a** $5.2 - 7 = -1.8$
        $-1.2 + 4 = 2.8$ Mid-point is at (−1.8, 2.8)

    b $5.2 \times 5 = 26$
        $-1.2 \times 5 = -6$ Mid-point is at (26, −6)

    c TE: $\begin{array}{l}(5.2 \times 5) - 7 = 19 \\ (-1.2 \times 5) + 4 = -2\end{array}$ Mid-point at (19, −2)

       ET: $\begin{array}{l}(5.2 - 7) \times 5 = -9 \\ (-1.2 + 4) \times 5 = 14\end{array}$ Mid-point at (−9, 14)

    TE and ET do not give the same image point; location is not independent of the order in which the transformations are carried out.

9 $\left(1 + \dfrac{p}{100}\right) \times 20\,000 + q = 40\,000$ gives $200p + q = 20\,000$ ...[1]

    $\left(1 + \dfrac{p}{100}\right) \times 7500 + q = 22\,500$ gives $75p + q = 15\,000$ ....[2]

    [1] − [2] gives $p = 40$ and $q = 12\,000$

    Total invested is $\dfrac{5 \times (33\,000 - 12\,000)}{1.40} = \$75\,000$.

    It appears to be unfair as the smaller the amount invested, the higher the percentage profit.

10 $1 \text{ psi} = 1 \text{ pound} \div 1 \text{ inch}^2 = 453.6 \text{ g} \div \dfrac{1}{0.3937^2} \text{ cm}^2 = 70.3078\ldots \text{ g/cm}^2$

    $\bar{x} = \dfrac{\Sigma x}{4} = 70.3078\ldots \bar{x} \text{ g/cm}^2$ gives $\Sigma x = 4 \times 70.3078\ldots \bar{x} = 281\bar{x} \text{ g/cm}^2$

EXERCISE 2E

1 a Median = $\frac{15+15}{2} = 15$

When an even number of values are ranked, the median is the mean of the two middle values.

b Median; it is greater than the mean (12.4) and a greater average suggests the dentist was busy.

c Failure to pay a bill or debt on time, for example.

A low average suggests low earnings.

2 a $\frac{11+12}{2} = 11.5$

$\Sigma f = 100$, so the median is the mean of 11 and 12 (the 50th and 51st values).

b The data are negatively skewed; $\bar{t} = \frac{1090}{100} = 10.9 <$ median.

3 a Median (29.5th value) = 6; mode = 8.

b Median is central to the values of x, but it occurs less frequently than all the others.

Mode is the most frequently occurring value, but it is also the highest value.

c $\Sigma f = 58$, so the 29th and 30th values must be 5 and 6.

Frequency for $x = 4$ must be increased from 14 to 16, so there are two incorrectly recorded values.

4 a Median (56th value) ≈ 4.4 minutes.

b Lower group median (28th) ≈ 2.8 minutes; upper group median (84th) ≈ 6.4 minutes.

5

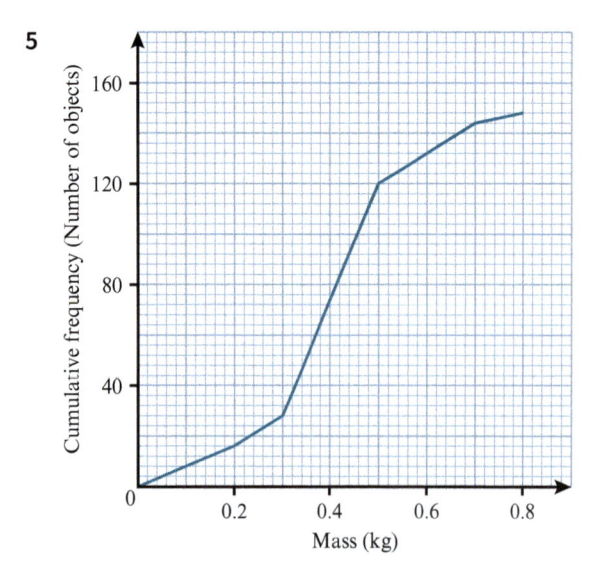

Median (74th value) ≈ 0.4 kg

a (cf-value at 0.5 kg) − (cf-value at 0.3 kg) is 120 − 28 = 92.

b Number of objects less than 0.2 kg or more than 0.6 kg is 16 + (148 − 132) = 32.

6   Mode remains as 15 and median remains as 16; mean decreases from 16 to 14.75.

7 a

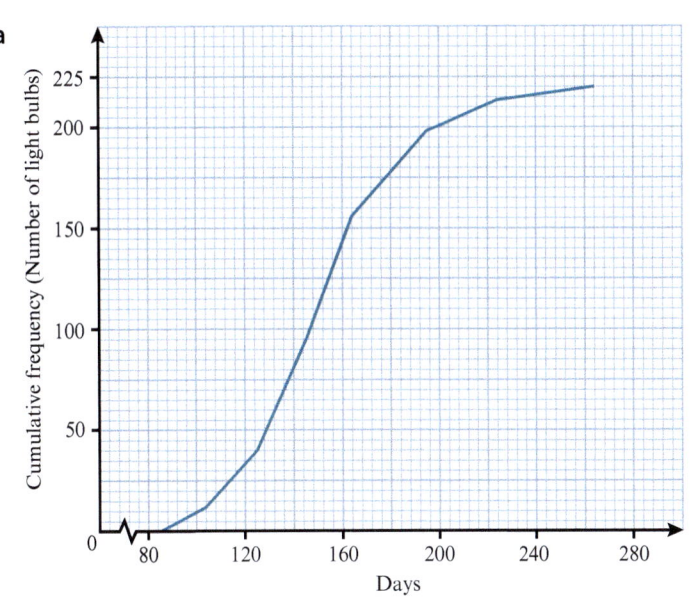

Polygon and curve both give a median (110th value) of approximately 150 days or 3600 hours.

   b To make the product appear longer-lasting, they are likely to use the highest average.

   Estimated mean is 152.84, which is greater than the median, so the mean appears to be advantageous.

   Upper boundary for the mean, which they might consider using, is:
   $$\frac{105 \times 12 + 125 \times 28 + 145 \times 54 + 165 \times 63 + 195 \times 41 + 225 \times 16 + 265 \times 6}{220} = 164.41 \text{ days}.$$

8   'Average' could refer to the mean, the median or the mode.

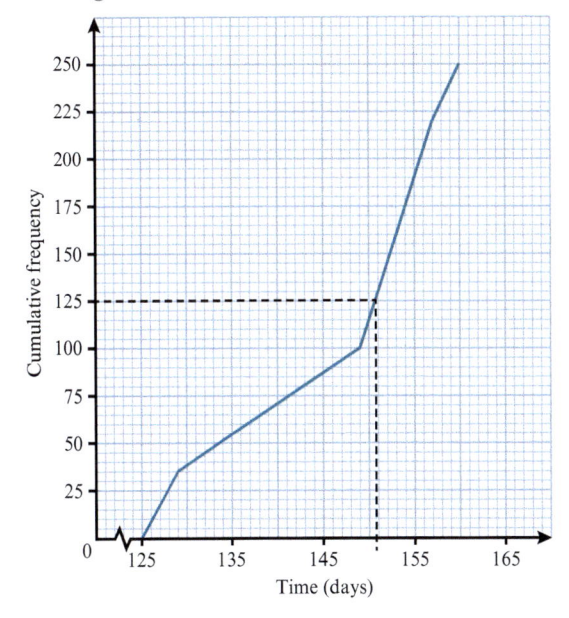

Median (125th value) ≈ 150.7 days or 3617 hours.

Mean ≈ $\frac{(127 \times 34) + (139 \times 66) + (153 \times 117) + (158.5 \times 33)}{250}$ = 146.5 days.

Modal class is $3576 \leq t < 3768$ hours or 149 to 157 days.

	Median	Mean	Modal class
Days	> 150	< 150	149 to 157
Supports claim	yes	no	?

The test is inconclusive and cannot be used to support or to refute the claim.

9 There is no mode; mean of $1 049 500 is distorted by the expensive home; median of $239 000 is the most useful.

10 a $p = 60 - 48 = 12$; $q = 60 - 20 = 40$; $r = 60 - 6 = 54$.

   b i

   This question can be answered without drawing the graphs shown here: the two tables have the same cumulative frequencies, one set descending and the other ascending.

   The reflection is in a horizontal line through a cumulative frequency value of 30.

   ii Median safe current = median unsafe current.

   The current referred to is approximately 1.82 amperes.

11 a First-half median is the 50th value, which is in the class $1 \leq t < 2$ minutes.

   Second-half median is the 50th value, which is in the class $4 \leq t < 5$ minutes.

   b i Second-half median < 5 minutes = 300 seconds.

   $k < \frac{300}{100}$, so the upper boundary value of k is 3.

   ii The first-half data are positively skewed, so mean is greater than median.

   The least possible first-half mean is:

   $\frac{0 \times 24 + 1 \times 38 + 2 \times 18 + 4 \times 12 + 5 \times 5 + 7 \times 3}{100} \times 60 = 100.8$, which is greater than the median of 100.

12 a

Score (S%)	Grade	f
$0 \leq S \leq 26$	E	15
$26 < S \leq 36$	D	20
$36 < S \leq 50$	C	25
$50 < S \leq 64$	B	15
$64 < S \leq 80$	A	5
		$\Sigma f = 80$

Score (S%)	cf
$S < 0$	0
$S \leq 26$	15
$S \leq 36$	35
$S \leq 50$	60
$S \leq 64$	75
$S \leq 80$	80

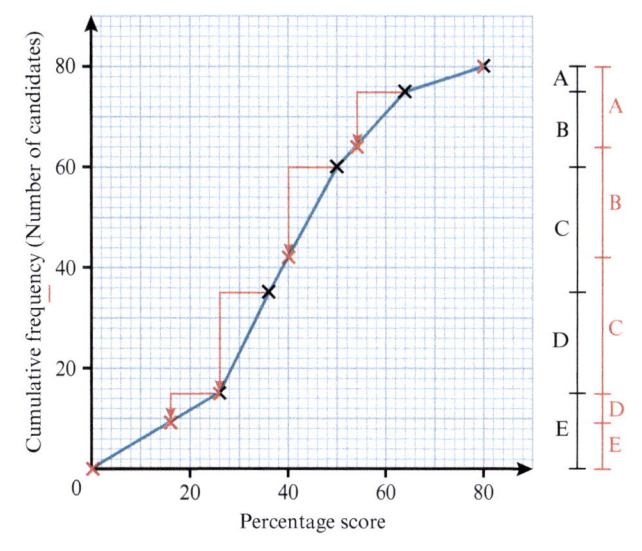

Integer cumulative frequencies of 0, 9, 15, 42, 64 and 80 can be used to plot the points for the new graph.

Median (40th value) $\approx 39\%$.

b

Grade	A	B	C	D	E
No. candidates (original boundaries)	5	15	25	20	15
No. candidates (reduced boundaries)	16	22	27	6	9
Change	+ 11	+ 7	+ 2	− 14	− 6

Total number of higher grades is $11 + 7 + 2 = 20$.

13 a i

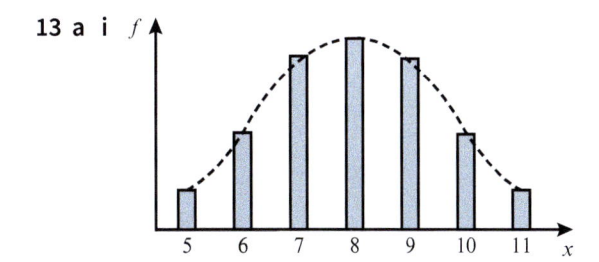

Only the symmetric curve is required; the bars are optional.

ii Mode = mean = median = 8

b Mode and median both remain as 8; mean increases from 8 to 9. The curve becomes positively skewed (longer tail at the right).

c $b = (42 \times 7) - (10 + 30 + 63 + 80 + 81 + 50 - 9) = 294 - 305 = -11$

Mode and median both remain as 8. The curve becomes negatively skewed (longer tail at the left).

> It is given that the mean decreases by 1 from 8 to 7.

14 Example curves:

> Any other symmetric curve (note that the mode can be the same as or different from the mean and median).

15 a Symmetrical. It suggests that the mean, median and mode are equal.

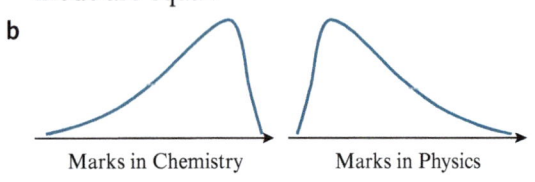

b

Marks in Chemistry   Marks in Physics

Chemistry: negatively skewed (longer tail at the left).

Physics: positively skewed (longer tail at the right).

Chapter 3
Measures of variation

EXERCISE 3A

1 a Range = 30 − 5 = 25; IQR = 25 − 8 = 17
 b Range = 37 − 2 = 35; IQR = 24.5 − 4.5 = 20
 c Range = 83 − 18 = 65; IQR = 64 − 39 = 25
 d Range = 113 − 17 = 96; IQR = 89 − 30 = 59
 e Range = 5.2 − − 3.3 = 8.5;
    IQR = 2.7 − − 2.9 = 5.6

2 a Range = 3.7 − 0.4 = 3.3;
    IQR = 2.9 − 1.15 = 1.75
 b Negative skew (longer tail at the left).

3 a Range = 50 − 9 = 41; IQR = 46 − 28 = 18
 b [box plot: Marks out of 50]
 c The box is symmetric, so the median is equal to the mean of the lower and upper quartiles.

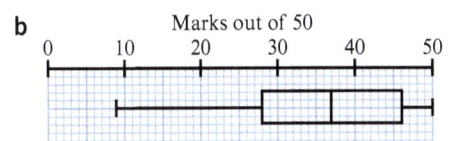

$$Q_2 = \frac{Q_1 + Q_3}{2}$$

$2Q_2 = Q_1 + Q_3$, which gives $Q_3 = 2Q_2 − Q_1$.

> We can easily check this, knowing that
> $Q_1 = 28$, $Q_2 = 37$ and $Q_3 = 46$: $46 = 2 \times 37 − 28$.

4 a Yes, if ranges alone are considered (both 13), but not if IQRs are considered (7 and 4).
 b [box plot: Number of fouls — hockey, football]

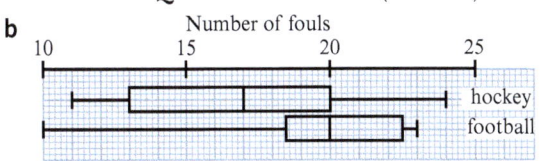

There were fewer fouls on average in hockey (medians are 17 and 20).

The numbers of fouls in hockey are more varied than in football.

5 a Ranges and IQRs are the same (35 and 18), but the two students' marks are quite different.
 b One of median (33 and 72) or mean (33 and 72), and one of the range or IQR.

6 a [cumulative frequency graph: cf (No. vintage cars) vs Maximum speed (km/h)]
 b Positive skew (longer tail at the right).

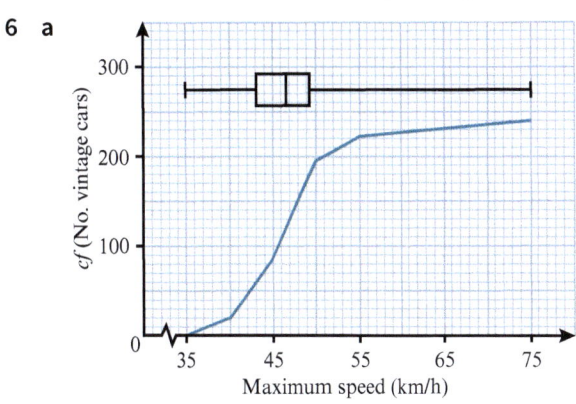

7 a i [box plots: Trips abroad — males, females]
 ii

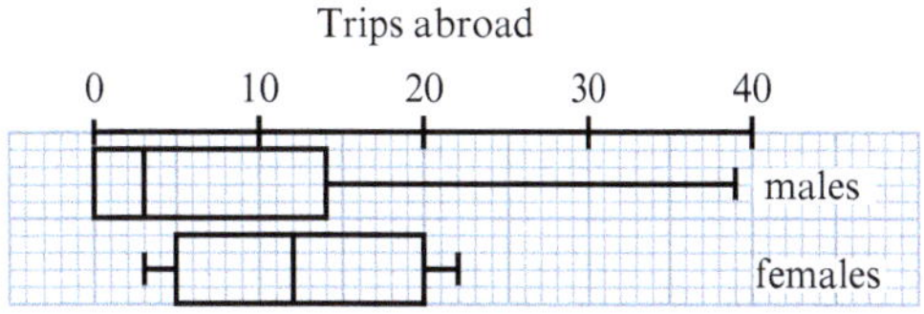

	Range	IQR	Median
Males	39	14	3
Females	19	15	12

On average, females have made more trips abroad than males.

Excluding the male who has made 39 trips, variation for males and females is similar.

 b The statement is not justified; there are no data about the number of different countries visited.

8 a 75th value − 25th value ≈ 0.225 − 0.095
      = 0.130 Ω
   b 90th value ≈ 0.345 Ω
   c ≈ 68th out of 100 values is the 68th percentile.
   d 70th value − 30th value ≈ 0.200 − 0.105
      = 0.095 Ω

   > The middle 40% is from the 30th to the 70th percentile of the 100 values.

9 a Width of polygon = 56 − 4 = 52 cm².
   b [box plot: Area (cm²), axis 0 to 60]
   c 144th value − 36th value ≈ 15.2 to 16.0 cm².

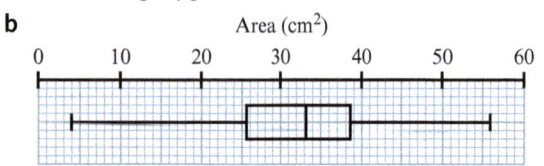

   > The middle 60% is from the 20th to the 80th percentile of the 180 values.

   d Outliers have areas less than 6.3 cm² or greater than 58.3 cm².
      Estimate ≈ 8, but there could be any number from 0 to 15 (any number of the 15 circuit boards with areas less than 8 cm² could have an area of less than 6.3 cm²).

10 a [Cumulative frequency graph (No. brackets) vs Deviation (d degrees)]

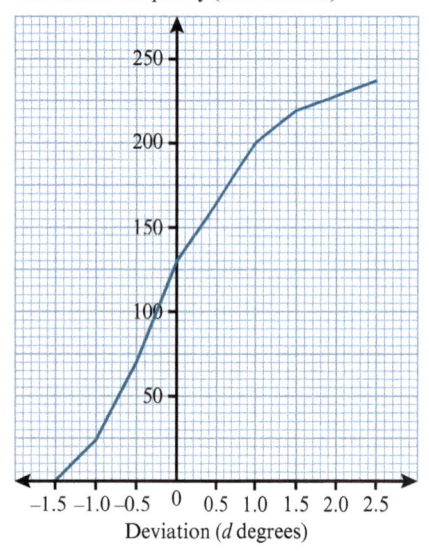

   b Q_1, Q_2 and Q_3 of the deviations are −0.6°, −0.1° and 0.7°.
      Q_1, Q_2 and Q_3 of the angles are 89.4°, 89.9° and 90.7°.
      Median angle is 89.9°; IQR for angles is 90.7 − 89.4 = 1.3°.
   c Percentage of brackets
      $\approx \dfrac{15 + (236 - 208)}{236} \times 100 = 18\%$

11 a $Q_3 - Q_1$ = 90th value − 30th value
      = 25 − 15 = 10
   b $0.85 \times 120 = 102$nd value = 30

12 a [Cumulative frequency graph: Number of 10g samples vs Mass (/ 0.01 g)]

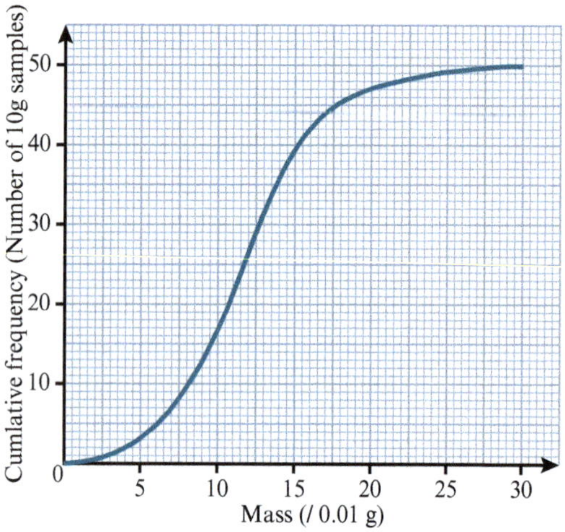

   b i $Q_3 - Q_1$ = 37.5th value − 12.5th value
         ≈ 0.145 − 0.0875 = 0.06 g
      ii 90th percentile − 10th percentile
         = 45th value − 5th value
         ≈ 0.1775 − 0.06 = 0.12 g
   c n = (cf-value at 22.5 g) − (cf-value at 7.5 g)
         ≈ 48 − 8 = 40
   d Variation is quite dramatic (from 0 up to a possible 3% of mass).
      Mushrooms are notoriously difficult to identify, so samples may not all be of the same type, and their toxicity varies by season.

13 The analysis should compare averages and variation (and skewness) and assess the effectiveness in reducing pollution levels for health benefits.

Chapter 3: Measures of variation

EXERCISE 3B

1 a Mean $= \dfrac{27+43+29+34+53+37+19+58}{8} = \dfrac{300}{8} = 37.5$

Mean of squares $= \dfrac{27^2+43^2+29^2+34^2+53^2+37^2+19^2+58^2}{8} = \dfrac{12478}{8}$

SD $= \sqrt{\dfrac{12478}{8} - \left(\dfrac{300}{8}\right)^2} = 12.4$

b Mean $= \dfrac{2.7}{6} = 0.45$; SD $= \sqrt{\dfrac{512.33}{6} - \left(\dfrac{2.7}{6}\right)^2} = 9.23$

2 a Var(B) $= \dfrac{21^2+33^2+45^2}{3} - \left(\dfrac{21+33+45}{3}\right)^2 = 96$

Var(C) $= \dfrac{41^2+53^2+65^2}{3} - \left(\dfrac{41+53+65}{3}\right)^2 = 96$

Var(P) $= \dfrac{51^2+63^2+75^2}{3} - \left(\dfrac{51+63+75}{3}\right)^2 = 96$

The mean for B is 33, so we could also find its variance using

$\dfrac{\Sigma(x-\bar{x})^2}{n} = \dfrac{(21-33)^2+(33-33)^2+(45-33)^2}{3}$
$= \dfrac{144+0+144}{3} = 96.$

b The three variances are identical.

The comments do not apply to Abraham's mean marks because they are all different:

B: $\dfrac{21+33+45}{3} = 33$; C: $\dfrac{41+53+65}{3} = 53$; P: $\dfrac{51+63+75}{3} = 63$

3 Mean $= \dfrac{59}{35} = 1\dfrac{24}{35}$ or 1.69; variance $= \dfrac{157}{35} - \left(\dfrac{59}{35}\right)^2 = 1.64$

Remember to use the exact mean (or its value to at least 4 significant figures) when finding the variance. If we use 1.69 instead, we will obtain an incorrect variance of 1.63.

4 a Mean $= \dfrac{720}{360} = 2$; SD $= \sqrt{\dfrac{1672}{360} - \left(\dfrac{720}{360}\right)^2} = 0.803$

b $Q_1 = Q_3 = 2$, so IQR $= 0$.

It tells us that the middle half (50%) of the values are identical, i.e. all equal to 2.

5 a Class mid-values are 25, 35, 50 and 70.

Girls: mean $= \dfrac{1200}{30} = 40$ minutes; SD $= \sqrt{\dfrac{53100}{30} - \left(\dfrac{1200}{30}\right)^2} = 13.0$ minutes

Boys: mean $= \dfrac{1600}{40} = 40$ minutes; SD $= \sqrt{\dfrac{74650}{40} - \left(\dfrac{1600}{40}\right)^2} = 16.3$ minutes

b i On average, the times spent were very similar, as both means are estimated to be 40 minutes.

ii Times spent by boys are more varied than times spent by girls.

6 Class mid-values are 16, 21, 27 and 33.5; SD $= \sqrt{\dfrac{28263}{50} - \left(\dfrac{1151}{50}\right)^2} = 5.94$ cm.

7 $$\frac{30k+16k+80+17k-51+180+152+60}{4k+23}=17$$
$$\frac{63k+421}{4k+23}=17$$
$63k+421=68k+391$, which gives $k=6$

$\text{Var}(x) = \frac{13711}{47} - 17^2 = 2.72$

8 **a** Class mid-values are 142, 147, 155 and 162.5.
$$\frac{142a+147b+(155\times 69)+(162.5\times 28)}{150}=153.14$$
$$142a+147b+15\,245=22\,971$$
$$142a+147b=7726 \quad \text{Q.E.D.}$$
$142a+147(53-a)=7726$
$-5a=-65$ gives $a=13$ and $b=40$.

See the 'Did You Know' feature in Section 4.3 of the Coursebook to find out the meaning and origins of Q.E.D.

b $\text{SD}=\sqrt{\frac{3\,523\,592}{150}-\left(\frac{22\,971}{150}\right)^2}=6.23$ cm

9 Total distance $=(k+1)+k+(k-3)+(k-8)+(k-15)+(k-24)+(k-35)=7k-84$.
Now, $7k-84=217$, so $k=43$; actual distances are 44, 43, 40, 35, 28, 19 and 8 km.
$\text{SD}=\sqrt{\frac{7819}{7}-\left(\frac{217}{7}\right)^2}=12.5$ km; IQR $=43-19=24$ km and IQR $\approx 2\times\text{SD}$.

10 **a** Class mid-values are 0.22, 0.58, 1.11 and 1.68.
$\text{Mean}\approx\frac{0.22\times 5+0.58\times 8+1.11\times 20+1.68\times 6}{39}=\frac{38.02}{39}=0.97$ tonnes

$\text{SD}\approx\sqrt{\frac{0.22^2\times 5+0.58^2\times 8+1.11^2\times 20+1.68^2\times 6}{39}-\left(\frac{38.02}{39}\right)^2}=\sqrt{\frac{44.5096}{39}-\left(\frac{38.02}{39}\right)^2}$
$=0.44$ tonnes

b

Mid-values	0	0.22	0.58	1.11	1.68
No. weeks (f)	13	5	8	20	6

$\text{Mean}\approx\frac{38.02}{52}=0.73$ tonnes; $\text{SD}\approx\sqrt{\frac{44.5096}{52}-\left(\frac{38.02}{52}\right)^2}=0.57$ tonnes

Mean decreases from 0.97 to 0.73 tonnes; SD increases from 0.44 to 0.57 tonnes.

11 Class mid-values are 27, 34.5, 42 and 53.
$$\frac{(27\times 14)+34.5x+42(30-x)+(53\times 6)}{50}\approx 37.32$$
$1956-7.5x=1866$ gives $x=12$, so $y=30-x=18$

One year later, Gudrun's age (g years) is included.
Class boundaries and mid-values are all greater by 1.

Ages are given in whole numbers of years, so only completed years are counted. A person who is 45.95 years old is said to be 45, not 46. Class boundaries are, therefore, 23, 31, 38, 46 and 60.

Age (years)	g	24–31	32–38	39–46	47–60
Mid-value	g	28	35.5	43	54
No. employees (f)	1	14	12	18	6

$$\frac{g+(28\times 14)+(35.5\times 12)+(43\times 18)+(54\times 6)}{51}=38$$

$$g+1916=1938$$

Gudrun's age is $g=22$ years.

Original variance $=\dfrac{73095}{50}-\left(\dfrac{1866}{50}\right)^2=69.12$ years²

New variance $=\dfrac{77361}{51}-\left(\dfrac{1938}{51}\right)^2=72.88$ years²

Variance increases (by 5.45%) from 69.12 to 72.88 years².

We must assume that none of the original 50 staff have been replaced.

12 **a** Distances in position 1: $\sqrt{125}, \sqrt{109}, \sqrt{101}, \sqrt{101}, \sqrt{109}, \sqrt{125}$.

Distances in position 2: $\sqrt{75}, \sqrt{79}, \sqrt{91}, \sqrt{111}, \sqrt{139}, \sqrt{175}$.

Change in mean is $\dfrac{\sqrt{75}+\sqrt{79}+\sqrt{91}+\sqrt{111}+\sqrt{139}+\sqrt{175}}{6}-\dfrac{2(\sqrt{125}+\sqrt{109}+\sqrt{101})}{6}=-0.116$

Mean decreases by 0.116 m or 11.6 cm.

Change in median is $\dfrac{\sqrt{91}+\sqrt{111}}{2}-\sqrt{109}=-0.403$

Median decreases by 0.403 m or 40.3 cm.

b Change in SD is $\sqrt{\dfrac{670}{6}-10.44...^2}-\sqrt{\dfrac{670}{6}-10.55...^2}=1.16$

SD increases by 1.16 m or 116 cm.

Change in IQR is $\sqrt{139}-\sqrt{79}-(\sqrt{125}-\sqrt{109})=2.16$

IQR increases by 2.16 m or 216 cm.

Change in range is $\sqrt{175}-\sqrt{75}-(\sqrt{125}-\sqrt{101})=3.44$

Range increases by 3.44 m or 344 cm.

c The discs get closer to P, but their distances from P become more varied.

The rotation causes a relatively small decrease in the average distance (1.1% for mean and 3.9% for median) but a relatively large increase in the variation of the distances (248% for SD, 292% for IQR and 304% for range).

d The squares of the distances from P are:

$(PA)^2 = 125 - 100\cos(90-\alpha)°$ $(PD)^2 = 101 - 20\cos(90+\alpha)°$
$(PB)^2 = 109 - 60\cos(90-\alpha)°$ $(PE)^2 = 109 - 60\cos(90+\alpha)°$
$(PC)^2 = 101 - 20\cos(90-\alpha)°$ $(PF)^2 = 125 - 100\cos(90+\alpha)°$

$\Sigma x^2 = (PA)^2 + (PB)^2 + (PC)^2 + (PD)^2 + (PE)^2 + (PF)^2$
$= 670 - 180\cos(90-\alpha)° - 180\cos(90+\alpha)°$
$= 670 - 180[\cos(90-\alpha)° + \cos(90+\alpha)°]$
$= 670$ Q.E.D.

> $(90-\alpha)°$ and $(90+\alpha)°$ are supplementary angles, so the sum of their cosines is zero.

EXERCISE 3C

1 a $\text{Var}(v) = \frac{\Sigma v^2}{n} - \left(\frac{\Sigma v}{n}\right)^2 = \frac{5480}{64} - \left(\frac{288}{64}\right)^2 = 65.375$

b $\text{SD}(w) = \sqrt{\frac{\Sigma w^2}{n} - \left(\frac{\Sigma w}{n}\right)^2} = \sqrt{\frac{4000}{36} - 5.2^2} = 9.17$

c $\frac{6120}{40} - \left(\frac{\Sigma xf}{40}\right)^2 = 12^2$

$\Sigma xf = 40 \times \sqrt{\frac{6120}{40} - 12^2} = 120$

d $\frac{\Sigma x^2 f}{50} - \left(\frac{2800}{50}\right)^2 = 100$, so $\Sigma x^2 f = 50 \times \left[100 + \left(\frac{2800}{50}\right)^2\right] = 161\,800$

e $\frac{193\,144}{n} - \left(\frac{2324}{n}\right)^2 = 3^2$

$\frac{193\,144}{n} - \frac{5400976}{n^2} = 9$

$9n^2 - 193144n + 5400976 = 0$

$(9n - 192892)(n - 28) = 0$, so $n = 28$

2 $\frac{8900}{n} - \left(\frac{220}{n}\right)^2 = 18^2$

$\frac{8900}{n} - \frac{48\,400}{n^2} = 324$

$324n^2 - 8900n + 48\,400 = 0$

$(324n - 2420)(n - 20) = 0$, so $n = 20$ and $\bar{x} = \frac{220}{n} = 11$

> The possible values of n can be found using the quadratic formula and, since n must be an integer, we can disregard non-integer values.

3 $\frac{\Sigma p^2 + \Sigma q^2}{25 + 25} - \left(\frac{\Sigma p + \Sigma q}{25 + 25}\right)^2 = \frac{6006 + 6114}{50} - \left(\frac{388 + 387}{50}\right)^2 = 2.15$

4 Mean $= \frac{14 \times 63.5 + 16 \times 57.3}{30} = 60.2$ kg; SD $= \sqrt{\frac{58444 + 56222}{30} - \left(\frac{1805.8}{30}\right)^2} = 14.1$ kg

5 a $\frac{26^2 + 29^2 + 30^2 + 34^2 + 26^2 + 24^2 + 27^2 + 31^2 + 30^2 + 28^2}{10} -$

$\left(\frac{26 + 29 + 30 + 34 + 26 + 24 + 27 + 31 + 30 + 28}{10}\right)^2 = 7.65$ Q.E.D.

b $\frac{7946 + 8199}{20} - \left(\frac{\Sigma x + 285}{20}\right)^2 = 31.6275$

$\left(\frac{\Sigma x + 285}{20}\right)^2 = 807.25 - 31.6275$

$\Sigma x = 20\sqrt{807.25 - 31.6275} - 285 = 272$

$\bar{x} = \frac{272}{10} = 27.2$ psi

6 a $\dfrac{\Sigma x^2 + \Sigma y^2}{n+29} - \left(\dfrac{\Sigma x + \Sigma y}{n+29}\right)^2 = \dfrac{7931 + \Sigma y^2}{n+29} - \left(\dfrac{397 + 499}{n+29}\right)^2 = \dfrac{7931 + \Sigma y^2}{n+29} - \dfrac{802\,816}{(n+29)^2} = 52$ Reduce the fractional equation by multiplying throughout by the lowest common denominator, which is $(n+29)^2$.

Now $(7931 + \Sigma y^2)(n+29) - 802\,816 = 52(n+29)^2$

$(n+29)\Sigma y^2 + 7931n + 229\,999 - 802\,816 = 52n^2 + 3016n + 43\,732$

$$\Sigma y^2 = \dfrac{52n^2 - 4915n + 616\,549}{n+29}$$

b $\dfrac{52n^2 - 4915n + 616\,549}{n+29} = 7941$

$52n^2 - 4915n + 616\,549 = 7941n + 230\,289$

$52n^2 - 12856n + 386\,260 = 0$

$(52n - 11036)(n - 35) = 0$ gives $n = 35$

7 Let the 6th value be a, then $250 + a = 6 \times 40$, so $a = -10$.

$\dfrac{\Sigma x^2}{5} - 50^2 = 15^2$, so $\Sigma x^2 = 5 \times (15^2 + 50^2) = 13\,625$

New variance $= \dfrac{13625 + (-10)^2}{6} - 40^2 = 687.5$

8 Let the winning score be x.

$\dfrac{2x^2 + 8 \times 34^2}{10} - \left(\dfrac{2x + 8 \times 34}{10}\right)^2 = 1.2^2$

$10(2x^2 + 9248) - (2x + 272)^2 = 144$

$x^2 - 68x + 1147 = 0$

$(x - 31)(x - 37) = 0$ gives $x = 31$ or $x = 37$

The winning score (which is the lowest) is 31.

9 Let the 15th book contain x pages.

$\Sigma(\text{pages}) = 3070 + x$, and $\Sigma(\text{pages}^2) = 685\,550 + x^2$.

$\dfrac{685\,550 + x^2}{15} - \left(\dfrac{3070 + x}{15}\right)^2 = 31.2^2$

$15(685\,550 + x^2) - (3070 + x)^2 = 219\,024$

$7x^2 - 3070x + 319\,663 = 0$

The quadratic formula gives $x = \dfrac{3070 \pm \sqrt{474336}}{14} = 170.1$ or 268.5.

Both solutions are valid, so the 15th book could contain 170, 268 or 269 pages.

10 Rearranging $\dfrac{\Sigma x^2}{N} - \bar{x}^2 = SD^2$ gives $\Sigma x^2 = N(SD^2 + \bar{x}^2)$.

For the $n + 2n = 3n$ values in the two datasets together:

Sum of squares $= n(S^2 + \bar{x}^2) + 2n\left(\dfrac{1}{4}S^2 + \bar{x}^2\right) = \dfrac{3n}{2}S^2 + 3n\bar{x}^2$.

Sum of values $= n\bar{x} + 2n\bar{x} = 3n\bar{x}$.

$SD = \sqrt{\dfrac{\frac{3n}{2}S^2 + 3n\bar{x}^2}{3n} - \left(\dfrac{3n\bar{x}}{3n}\right)^2} = \sqrt{\dfrac{1}{2}S^2 + \bar{x}^2 - \bar{x}^2} = \dfrac{S}{\sqrt{2}}$ or $\dfrac{S\sqrt{2}}{2}$

EXERCISE 3D

1 SD(M) = 8 kg (unaffected by addition of 5);
SD(W) = 6 kg (unaffected by addition of −3).

2 $SD(y) = SD(y-5) = \sqrt{\frac{890}{20} - \left(\frac{130}{20}\right)^2} = 1.5$

3 $\text{Mean}(r) = \frac{\Sigma(r-3)}{365} + 3 = \frac{1795.8}{365} + 3 = 7.92 \text{ mm}$

$\text{Var}(r) = \text{Var}(r-3)$

$\frac{\Sigma r^2}{365} - 7.92^2 = \frac{9950}{365} - 4.92^2$

$\Sigma r^2 = 365 \times \left(\frac{9950}{365} - 4.92^2 + 7.92^2\right) = 24\,009.8$

4 Variance today = Variance 20 years ago

$\frac{24\,224}{n} - (15.7 + 20)^2 = \frac{16\,000}{n} - 15.7^2$

$\frac{8224}{n} = 1028$ gives $n = 8$

5 $\text{Var}(y-3) = \text{Var}(y)$

$\frac{2775}{n} - \left(\frac{105 - 3n}{n}\right)^2 = 13^2$

$2775n - (105 - 3n)^2 = 169n^2$

$178n^2 - 3405n + 11025 = 0$

$(178n - 735)(n - 15) = 0$ gives $n = 15$

6 Each classmate's true height is 1.2 cm greater than the measurement that was taken.

The mean of 163.8 cm is not valid (true mean is 163.8 + 1.2 = 165 cm).

SD of 7.6 cm is valid (all true heights were reduced by 1.2 cm, which does not affect the variation).

7 All journey times are reduced by 15 minutes, so the mean would be 4 hours 20 minutes.

SD is not affected and remains as $\sqrt{53.29} = 7.3$ minutes.

This might be possible if the changes result in the coaches avoiding busy traffic conditions.

8 $\text{Var}(x-4) = \text{Var}(x) = \frac{12x^2 + 8(x-2)^2}{20} - \left(\frac{12x + 8(x-2)}{20}\right)^2$

$= \frac{20x^2 - 32x + 32}{20} - \frac{400x^2 - 640x + 256}{400}$

$= x^2 - 1.6x + 1.6 - x^2 + 1.6x - 0.64$

$= 0.96$, so the variance of the reduced lengths is 0.96 cm^2.

Each pair of jeans/pants consists of two leg lengths. A more basic approach is to find the variance of twelve '−4s' and eight '−6s'. (In fact, any two numbers that differ by 2 with frequencies in the ratio 3 : 2 has the same variance.)

9 a Mean = $\frac{56}{7} = 8$; SD = $\sqrt{\frac{560}{7} - \left(\frac{56}{7}\right)^2} = 4$

b Mean(odd) = mean(even) − 1 = 7; SD(odd) = SD(even) = 4.

c

n	2	3	4	5	6	7	8	9	10
Variance (V)	1	$\frac{8}{3}$	5	8	$\frac{35}{3}$	16	21	$\frac{80}{3}$	33
3V	3	8	15	24	35	48	63	80	99
3V + 1	4	9	16	25	36	49	64	81	100
$\sqrt{3V+1}$	2	3	4	5	6	7	8	9	10

$\sqrt{3V+1} = n$, so $V = \frac{n^2 - 1}{3}$. The nth term is $\frac{n^2 - 1}{3}$.

$\frac{n^2 - 1}{3}$ is also the nth term for the variance of the first n positive odd integers.

10 a $\Sigma u + \Sigma c = 9 + (15 \times 1) + 19 = 43$ goals

b $\text{Var}(u) = \text{Var}(u-1)$
$$\frac{\Sigma u^2}{15} - \left(1\frac{9}{15}\right)^2 = \frac{25}{15} - \left(\frac{9}{15}\right)^2$$
$$\Sigma u^2 = 15 \times \left(\frac{25}{15} - \left(\frac{9}{15}\right)^2 + \left(1\frac{9}{15}\right)^2\right) = 58$$

c $\text{Var}(u \text{ and } c) = \frac{\Sigma u^2 + \Sigma c^2}{15 + 15} - \left(\frac{\Sigma u + \Sigma c}{15 + 15}\right)^2 = \frac{58 + 39}{30} - \left(\frac{24 + 19}{30}\right)^2 = 1.179$

11 a $\bar{x} = \frac{44}{20} + 1 = 3.2$, so $\Sigma x = 20 \times 3.2 = 64$ Q.E.D.

$\text{Var}(x) = \text{Var}(x-1)$
$$\frac{\Sigma x^2}{20} - 3.2^2 = \frac{132}{20} - \left(\frac{44}{20}\right)^2$$
$$\Sigma x^2 = 20 \times \left(\frac{132}{20} - \left(\frac{44}{20}\right)^2 + 3.2^2\right) = 240$$

b $\bar{y} = \frac{1184}{80} - 1 = 13.8$, so $\Sigma y = 80 \times 13.8 = 1104$

$\text{Var}(y) = \text{Var}(y+1)$
$$\frac{\Sigma y^2}{80} - 13.8^2 = \frac{17704}{80} - \left(\frac{1184}{80}\right)^2$$
$$\Sigma y^2 = 80 \times \left(\frac{17704}{80} - \left(\frac{1184}{80}\right)^2 + 13.8^2\right) = 15\,416$$

c $\text{Var}(x \text{ and } y) = \frac{\Sigma x^2 + \Sigma y^2}{80 + 20} - \left(\frac{\Sigma x + \Sigma y}{80 + 20}\right)^2 = \frac{240 + 15\,416}{100} - \left(\frac{64 + 1104}{100}\right)^2 = 20.1376$

12 a $\bar{x} = \frac{1820}{200} + 160 = 169.1$, and $\Sigma x = 200 \times 169.1 = 33\,820$

$\bar{y} = \frac{2250}{300} + 150 = 157.5$, and $\Sigma y = 300 \times 157.5 = 47\,250$

Mean height is $\frac{\Sigma x + \Sigma y}{200 + 300} = \frac{33820 + 47250}{500} = 162.14$ cm

b $\text{Var}(x) = \text{Var}(x - 160)$
$$\frac{\Sigma x^2}{200} - 169.1^2 = \frac{18240}{200} - \left(\frac{1820}{200}\right)^2$$
$$\Sigma x^2 = 200 \times \left(\frac{18240}{200} - \left(\frac{1820}{200}\right)^2 + 169.1^2\right) = 5\,720\,640$$

$\text{Var}(y) = \text{Var}(y - 150)$
$$\frac{\Sigma y^2}{300} - 157.5^2 = \frac{20100}{300} - \left(\frac{2250}{300}\right)^2$$
$$\Sigma y^2 = 300 \times \left(\frac{20100}{300} - \left(\frac{2250}{300}\right)^2 + 157.5\right) = 7\,445\,100$$

$\text{Var}(x \text{ and } y) = \frac{\Sigma x^2 + \Sigma y^2}{200 + 300} - 162.14^2 = \frac{5720640 + 7445100}{500} - 162.14^2 = 42.1004$ cm^2

EXERCISE 3E

1 $0.80 \times 0.80 = \$0.64$

2 $SD(x) = \frac{1}{2} \times SD(2x) = \frac{1}{2} \times \sqrt{\frac{14600}{20} - \left(\frac{420}{20}\right)^2} = 8.5$

Alternatively, $\Sigma x^2 = \frac{14600}{4} = 3650$, and $\Sigma x = \frac{420}{2} = 210$, so

$SD(x) = \sqrt{\frac{3650}{20} - \left(\frac{210}{20}\right)^2} = 8.5$

3 $y = 3x + 1$, so $SD(y) = 3 \times SD(x) = 3 \times 0.88 = 2.64$

4 a $\Sigma 10(T-30) = 10 \times (\Sigma T - 7 \times 30) = 10 \times (223.3 - 210) = 133$ or
$\Sigma 10(T-30) = 10 \times (2.1 + 1.7 + 1.2 + 1.5 + 1.9 + 2.2 + 2.7) = 133$
$\Sigma 100(T-30)^2 = 100 \times (\Sigma T^2 - 60\Sigma T + 7 \times 900)$
$= 100 \times (7124.73 - 13\,398 + 6300) = 2673$ or
$\Sigma 100(T-30)^2 = 100 \times (2.1^2 + 1.7^2 + 1.2^2 + 1.5^2 + 1.9^2 + 2.2^2 + 2.7^2) = 2673$

b $\frac{1}{10}\sqrt{\frac{2673}{7} - \left(\frac{133}{7}\right)^2} = 0.457°C$

Both answers from part **a** must be used in this calculation.

c $(0.45669...)^2 = 0.209(°C)^2$

5 $315 \times 240 = \$75\,600$

6 a $1.8 \times 15 = 27°F$

b Mean $= \frac{54.5 - 32}{1.8} = 12.5°C$; SD $= \frac{\sqrt{65.61}}{1.8} = 4.5°C$

7 a Fruit & veg; mean is unchanged, so the total is unchanged.

b Tinned food; mean increased, but standard deviation was unchanged.

c Bakery; mean and standard deviation both decreased by 10%.

8 $\Sigma(1.3x)^2 = 1.69\Sigma x^2 = 0.0507$ km²

$\Sigma x^2 = \frac{0.0507}{1.69} = 0.03$ km² or 30000 m²

$Var(x) = Var(x - 20)$

$\frac{30\,000}{45} - \bar{x}^2 = \frac{1200}{45} - (\bar{x} - 20)^2$

$\frac{28\,800}{45} - \bar{x}^2 = -\bar{x}^2 + 40\bar{x} - 400$

$40\bar{x} = 1040$ gives $\bar{x} = 26$, so the mean natural length of the ropes is 26 metres.

9 Let the original rate be £1 = €x, so €1 = £$\frac{1}{x}$.

New rate is £1 = €$(1 - 0.1525)x$ = € $0.8475x$, so €1 = £$\frac{1}{0.8475x}$.

Alternatively, $\frac{0.1525}{1 - 0.1525} \times 100$
$= 18.0\%$ increase.

% increase in euro value is $\frac{\frac{1}{0.8475x} - \frac{1}{x}}{\frac{1}{x}} \times 100 = \left(\frac{100}{0.8475} - 100\right) = 18.0\%$

Chapter 4
Probability

EXERCISE 4A

1 a Each of the 12 boys and 24 girls is equally likely to be selected, so
$$P(\text{a particular boy}) = \frac{1}{12+24} = \frac{1}{36}$$
b Of the 36 students, 24 are girls, so
$$P(\text{girl}) = \frac{24}{36} = \frac{2}{3}.$$

2 a From the team's previous results.
b $P(\text{lose}) = 1 - [P(\text{win}) + P(\text{draw})]$
$= 1 - [0.65 + (1 - 0.85)] = 0.2$
Expectation $= 40 \times P(\text{lose}) = 40 \times 0.2$
$= 8$ games
c They may win some of the games that they are expected to draw.

3 Three of the ten cards are red with A or C.
Expectation $= n \times P(\text{red with A or C})$
$= 40 \times \frac{3}{10} = 12$

4 a Expectation $= n \times P(\text{not } 4) = 400 \times \frac{6}{8} = 300$
b $(400 + x) \times \frac{2}{8} \geq 160$ gives $x \geq 240$, so he must spin the wheel at least 240 more times.

5 a The smallest possible number of counters in the bag is six, one of which is black. So the smallest possible number of white counters is five.
b Let the bag contain B black counters and W white counters, then $\frac{B}{B+W} = \frac{1}{6}$.
Substituting $B = 3$ gives $W = 15$: the smallest possible number of white counters is 15.

6 $P(\text{selecting a particular coin}) = 1 - 0.98 = 0.02$
$= \frac{1}{50} = \frac{1}{\text{number of coins}}$
There are 50 coins in the box.

7 Mean $= \dfrac{8+13+17+18+24+32+34+38}{8} = 23$

SD $= \sqrt{\dfrac{8^2+13^2+17^2+18^2+24^2+32^2+34^2+38^2}{8} - 23^2} \approx 10.1$

Three of the eight values are outside the range from 12.9 to 33.1.
$P(\text{more than one SD from mean}) = \dfrac{3}{8}$

8 Let the number of girls be G, then $\dfrac{G}{G+837} = \dfrac{4}{7}$, giving $G = 1116$.
There are $837 + 1116 = 1953$ students altogether, and $P(\text{a particular boy}) = \dfrac{1}{1953}$.

EXERCISE 4B

1 a The events are mutually exclusive:

$$\frac{3}{6} + \frac{1}{6} - 0 = \frac{2}{3}$$

> A number rolled with a die cannot be prime and also a 4.

b The events are mutually exclusive:

$$\frac{2}{6} + \frac{2}{6} - 0 = \frac{2}{3}$$

c The events are not mutually exclusive:

$$\frac{3}{6} + \frac{3}{6} - \frac{1}{6} = \frac{5}{6}$$

> There is one number on the die (4) that is more than 3 and also a factor of 8.

2 a Girls who took the test.

b The events are not mutually exclusive:

$$\frac{19}{40} + \frac{7}{40} - \frac{3}{40} = \frac{23}{40} \text{ or } \frac{16 + 3 + 4}{40} = \frac{23}{40}$$

3 a i The events are not mutually exclusive:

$$\frac{8}{55} + \frac{30}{55} - \frac{5}{55} = \frac{3}{5} \text{ or } \frac{5 + 3 + 25}{55} = \frac{3}{5}$$

ii The events are not mutually exclusive:

$$\frac{25}{55} + \frac{47}{55} - \frac{22}{55} = \frac{10}{11} \text{ or } \frac{3 + 22 + 25}{55} = \frac{10}{11}$$

b Male or a goat ≡ not a female sheep.

A sheep or female ≡ not a male goat.

4 a i (3, 3), because the sum is 6 and the difference is 0.

ii (2, 4), (4, 2), because both have sums of 6 and both have two even numbers.

iii (2, 2), (4, 4), (6, 6), because all have differences of 0 and all have two even numbers.

b Each pair of events has at least one common favourable outcome, so X, Y and Z are not mutually exclusive.

5 a The events are not mutually exclusive:

$$\frac{2}{8} + \frac{3}{8} - \frac{1}{8} = \frac{1}{2}$$

b The events are not mutually exclusive:

$$\frac{6}{8} + \frac{6}{8} - \frac{5}{8} = \frac{7}{8}$$

6 a Solve $a + b + c + 10 = 25$; $a + b = 9$ and $b + c = 8$.
$a = 7$, $b = 2$ and $c = 6$.

b i The events are not mutually exclusive:

$$\frac{9}{25} + \frac{8}{25} - \frac{2}{25} = \frac{3}{5} \text{ or } \frac{7 + 2 + 6}{25} = \frac{3}{5}$$

ii $\frac{7 + 6}{25} = \frac{13}{25}$

7 a

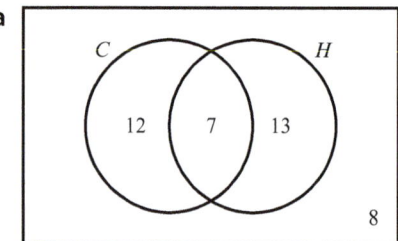

b i $\frac{12}{40} = \frac{3}{10}$ **ii** $\frac{12 + 13}{40} = \frac{5}{8}$

8 $100 - (28 + 8 + 20) = 44\%$

9 a $\frac{13 + 47 + 5 + 18 + 11 + 26}{132} = \frac{10}{11}$ or

$$\frac{132 - 12}{132} = \frac{10}{11}$$

b $\frac{12 + 13 + 47}{132} = \frac{6}{11}$ or

$$\frac{132 - (11 + 5 + 18 + 26)}{132} = \frac{6}{11}$$

c $\frac{13 + 11 + 18}{132} = \frac{7}{22}$ or

$$\frac{132 - (12 + 47 + 26 + 5)}{132} = \frac{7}{22}$$

10 a Students who study Pure Mathematics and Statistics, but not Mechanics.

b i $\dfrac{82+39-32}{100} = \dfrac{89}{100}$ or $\dfrac{33+17+32+7}{100} = \dfrac{89}{100}$

ii $\dfrac{17+7}{100} = \dfrac{6}{25}$

c Mechanics, Statistics, Pure Mathematics.

11 a X and Y are not mutually exclusive because $P(X \cap Y) \neq 0$.

> Also, X and Y are not mutually exclusive because $X \cap Y \neq \emptyset$ and because $P(X \cup Y) \neq P(X) + P(Y)$.

b $0.5 + 0.6 - 0.2 = 0.9$

c $(0.5 - 0.2) + (0.6 - 0.2) = 0.7$

12 a A and C.

> A and C are mutually exclusive because $P(A \cap C) = 0$, $A \cap C = \emptyset$ and because $P(A \cup C) = P(A) + P(C)$.

b $1 - (0.18 + 0.12 + 0.18 + 0.1 + 0.2) = 0.22$

13 a This is the area of the board where the cards overlap, i.e. their intersection. $\dfrac{4 \times 6}{30 \times 30} = \dfrac{2}{75}$

b This is the area of the board covered by either card, i.e. their union.
$\dfrac{(8 \times 12) + (15 \times 20) - (4 \times 6)}{30 \times 30} = \dfrac{31}{75}$

c This is 'the union – the intersection':
$\dfrac{31}{75} - \dfrac{2}{75} = \dfrac{29}{75}$

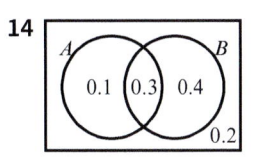

14
a P(in A or not in B) $= (0.1 + 0.3) + 0.2 = 0.6$

b P(in B but not in A) $= 0.4$

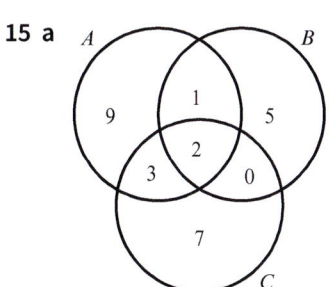

15 a

b $9 + 3 + 7 = 19$; they had not visited Burundi.

c They had visited Angola or Burundi, but not Cameroon; $9 + 1 + 5 = 15$.

d $\dfrac{1+2+3+0}{27} = \dfrac{2}{9}$ or $\dfrac{27-(9+5+7)}{27} = \dfrac{2}{9}$

EXERCISE 4C

1

```
        0.5 ── H ...... P(HH) = 0.5 × 0.5
   0.5 ─ H
        0.5 ── T ...... P(HT) = 0.5 × 0.5
   0.5 ─ T
        0.5 ── H ...... P(TH) = 0.5 × 0.5
        0.5 ── T ...... P(TT) = 0.5 × 0.5
```

$P(HT) + P(TH) = (0.5 \times 0.5) + (0.5 \times 0.5)$
$\qquad = 0.5$ or $\dfrac{1}{2}$

2 The grid shows products.

6	6	12	18	24	30	36
5	5	10	15	20	25	30
4	4	8	12	16	20	24
3	3	6	9	12	15	18
2	2	4	6	8	10	12
1	1	2	3	4	5	6
	1	2	3	4	5	6

> A grid such as this is called a *possibility diagram*, *outcome space* or *sample space*.
>
> Probabilities can be determined by counting to find what proportion of the 36 equally-likely outcomes is favourable to a particular event. No further working is required.

a One of the 36 equally-likely outcomes is favourable: $\frac{1}{36}$

b Nine of the 36 equally-likely outcomes are favourable: $\frac{9}{36} = \frac{1}{4}$

c Four of the 36 equally-likely outcomes are favourable: $\frac{4}{36} = \frac{1}{9}$

3 a P(mech.) × P(elec.) = 0.08 × 0.15 = 0.012
  b P(not mech.) × P(not elec.) = (1 − 0.08) × (1 − 0.15) = 0.782

4 P(win, not win) + P(not win, win) = (0.7 × 0.3) + (0.3 × 0.7) = 0.42

5 a P(W, W') + P(W', W) + P(W, W) = 0.6 × 0.4 + 0.4 × 0.6 + 0.6 × 0.6 = 0.84

> 'At least one' has the same meaning as 'not none', so we could use
> 1 − P(W', W') = 1 − 0.4 × 0.4 = 0.84.

  b 1 − [P(L, L) + P(L, D) + P(D, L)] = 1 − [(0.3 × 0.3) + (0.3 × 0.1) + (0.1 × 0.3)] = 0.85

6 a P(S') = 1 − P(S) = 1 − 0.3 = 0.7
    P(S', S', S') = 0.7 × 0.7 × 0.7 = 0.343
  b P(S, S', S') + P(S', S, S') + P(S', S', S) = 3 × (0.3 × 0.7 × 0.7) = 0.441

> The one day on which it snows is equally likely to be the first, second or third day.

7 a i 0.85 × 0.64 = 0.544
    ii 0.85 × (1 − 0.4) × 0.64 = 0.3264
    iii (0.85 × 0.4 × 0.36) + (0.85 × 0.6 × 0.64) + (0.15 × 0.4 × 0.64) = 0.4872
  b It means that the outcome of any of these three sporting events has no effect on the probabilities of the outcomes of the other two sporting events.

    This may not be true because, for example, winning one event may increase an athlete's confidence, so that they are more likely to win another event.

8 a

Values of S

Q							
	−2	−2	−1	0	0	1	2
	−1	−1	0	1	1	2	3
	−1	−1	0	1	1	2	3
	0	0	1	2	2	3	4
		0	1	2	2	3	4

P

Values of S^2

Q							
	−2	4	1	0	0	1	4
	−1	1	0	1	1	4	9
	−1	1	0	1	1	4	9
	0	0	1	4	4	9	16
		0	1	2	2	3	4

P

$P(S = 2) = \dfrac{5}{24}$

b $P(S^2 = 1) = \dfrac{9}{24} = \dfrac{3}{8}$

9 a The statement is not true. Any number from none to ten may be delivered; nine is just the average.
 b P(all three do not arrive after 1 day) = $0.5 \times 0.5 \times 0.5 = 0.125$
 c P(letter same day, package after 1 day) + P(letter after 1 day, package after 2 days)
  = $(0.4 \times 0.55) + (0.5 \times 0.3) = 0.37$

10 a Each square block in the histogram represents 5% (or 0.05) of the buses.
 $0.6 \times 0.6 = 0.36$ or $\dfrac{9}{25}$

 b P(at least one > 7) = 1 − P(both ⩽ 7) = $1 - (0.85 \times 0.85) = \dfrac{111}{400}$ or 0.2775

11 a A represents 'answers the phone'.

 P(A) + P(A′, A) = $0.6 + (0.4 \times 0.6) = 0.84$.

 b P(A) + P(A′, A) + P(A′, A′, A) + P(A′, A′, A′, A)
  = $0.6 + (0.4 \times 0.6) + (0.4 \times 0.4 \times 0.6)$
  + $(0.4 \times 0.4 \times 0.4 \times 0.6) = 0.9744$

 Alternatively,
 1 − P(A′, A′, A′, A′) = $1 - 0.4^4 = 0.9744$

12 a P(one particular newspaper on two consecutive mornings) = $\dfrac{1}{4} \times \dfrac{1}{4}$, so

 P(the same newspaper on two consecutive mornings) = $4 \times \dfrac{1}{4} \times \dfrac{1}{4} = \dfrac{1}{4}$

 b Any of four newspapers on the first morning, any of three on the second morning and any of two on the third morning: $\left(4 \times \dfrac{1}{4}\right) \times \left(3 \times \dfrac{1}{4}\right) \times \left(2 \times \dfrac{1}{4}\right) = \dfrac{24}{64} = \dfrac{3}{8}$.

13 P(head with each toss) = $\sqrt[3]{\dfrac{125}{512}} = \dfrac{5}{8}$, so P(no head with 3 tosses) = $\left(1 - \dfrac{5}{8}\right)^3 = \dfrac{27}{512}$.

14 a P(male teacher) × P(female teacher) = $\frac{2}{5} \times \frac{1}{4} = \frac{1}{10}$ or 0.1

b $3 \times \frac{1}{5} \times \frac{1}{4} = \frac{3}{20}$ or 0.15

c $\frac{3}{5} \times \frac{2}{4} = \frac{3}{10}$ or 0.3

15 a i Substituting $x = 5$ gives $P(5) = \frac{k-5}{25}$.

ii P(score < 3) = P(1 or 2) = $\frac{k-2}{25} + \frac{k-1}{25} = \frac{2k-3}{25}$

b Sum of probabilities is $\frac{k-1+k-2+k-3+k-4+k-5}{25} = 1$, giving $5k - 15 = 25$, so $k = 8$.

P(sum < 5) = P(1, 1, 1) + P(1, 1, 2) + P(1, 2, 1) + P(2, 1, 1)

$= \left(\frac{7}{25}\right)^3 + 3 \times \left(\frac{7}{25} \times \frac{7}{25} \times \frac{6}{25}\right) = \frac{49}{625}$ or 0.0784

16 a i P(on start) = $\frac{1}{6}$ by rolling 2

ii P(on start) = $\frac{8}{36} = \frac{2}{9}$ by rolling (1, 1), (1, 6), (2, 2), (3, 4), (4, 3), (5, 2), (6, 1) or (6, 6)

b i 18 is scored by rolling three 6s, but this leaves the counter on square 6: P(on 18) = 0

ii P(on 17) = $\frac{2}{216} = \frac{1}{108}$ by rolling (5, 6, 6) or (6, 5, 6)

> By rolling (6, 6, 5), a counter ends on square 5.

EXERCISE 4D

1 $P(Y \cap Z) = P(Y) \times P(Z) = 0.7 \times 0.9 = 0.63$

2 $P(N) = \frac{P(M \cap N)}{P(M)} = \frac{0.21}{0.75} = 0.28$

3 a $P(S \cap T) = P(S) \times P(T) = 0.4 \times (1 - 0.2) = 0.32$

b $P(S' \cap T) = P(S') \times P(T)$
$= (1 - 0.4) \times (1 - 0.2) = 0.48$

4 a i $P(A) = \frac{0.35}{P(B)}$ **ii** $P(A) = \frac{0.4}{P(C)}$

b i $P(B \cap C) = P(B) \times P(C) = 0.56$, so

$P(B) = \frac{0.56}{P(C)}$.

$P(C) = \frac{0.4 \times P(B)}{0.35}$ gives $P(B) = \frac{0.56 \times 0.35}{0.4 \times P(B)}$.

$[P(B)]^2 = \frac{0.56 \times 0.35}{0.4} = 0.49$, so $P(B) = 0.7$

ii $P(A') = 1 - 0.5 = 0.5$

iii $P(B' \cap C') = P(B') \times P(C')$
$= (1 - 0.7) \times (1 - 0.8) = 0.06$

Chapter 4: Probability

5 a

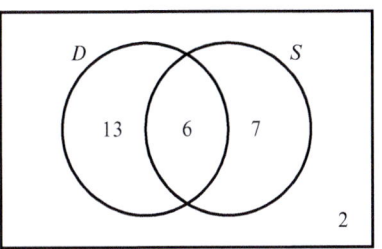

	S	S'	Totals
D	6	13	19
D'	7	2	9
Totals	13	15	28

Venn diagram: Total 28. D and S overlap with values 13, 6, 7, and 2 outside.

b $P(D \cap S) = \frac{6}{28}$, $P(D) = \frac{19}{28}$, $P(S) = \frac{13}{28}$ and
$\frac{6}{28} \neq \frac{19}{28} \times \frac{13}{28} = \frac{247}{784}$
$P(D \cap S) \neq P(D) \times P(S)$, so D and S are not independent.

6 $P(R \cap M) = \frac{20}{80}$, $P(R) = \frac{32}{80}$, $P(M) = \frac{50}{80}$ and
$\frac{20}{80} = \frac{32}{80} \times \frac{50}{80}$
$P(R \cap M) = P(R) \times P(M)$, so R and M are independent.

7 a $P(A) = \frac{9}{16}$, $P(B) = \frac{3}{4}$, $P(A \cap B) = \frac{1}{2}$

b $\frac{1}{2} \neq \frac{9}{16} \times \frac{3}{4} = \frac{27}{64}$, so $P(A \cap B) \neq P(A) \times P(B)$; A and B are not independent.

c A and B both occur when, for example, 1 and 2 are rolled, i.e. $P(A \cap B) \neq 0$.

8 a $P(X \text{ and } Y) = \frac{1}{12}$, $P(X) = \frac{1}{4}$, $P(Y) = \frac{1}{3}$
and $\frac{1}{12} = \frac{1}{4} \times \frac{1}{3}$.
$P(X \cap Y) = P(X) \times P(Y)$, so X and Y are independent.

b X and Y are not mutually exclusive; X and Y both occur when, for example, 1 and 5 are rolled, i.e. $P(X \cap Y) \neq 0$.

9 $P(V \cap W) = \frac{1}{16}$, $P(V) = \frac{1}{8}$, $P(W) = \frac{27}{64}$ and
$\frac{1}{16} \neq \frac{1}{8} \times \frac{27}{64} = \frac{27}{512}$
$P(V \cap W) \neq P(V) \times P(W)$, so V and W are not independent.

10 a

	B	B'	Totals
M	60	48	108
M'	50	42	92
Totals	110	90	200

b Ownership is not independent of gender.

For M and B: $\frac{60}{200} \neq \frac{108}{200} \times \frac{110}{200}$

For M' and B: $\frac{50}{200} \neq \frac{92}{200} \times \frac{110}{200}$

For M and B': $\frac{48}{200} \neq \frac{108}{200} \times \frac{90}{200}$

For M' and B': $\frac{42}{200} \neq \frac{92}{200} \times \frac{90}{200}$

Only one of these four calculations is actually required.

c Females $\frac{50}{92} \times 100 = 54.3\%$; males $\frac{60}{108} \times 100 = 55.6\%$

If ownership and gender were independent then these percentages would be equal.

11 $a = \frac{3100 \times 7440}{12\,400} = 1860$; $b = \frac{6280 \times 7440}{12\,400} = 4092$; $c = \frac{2480 \times 7440}{12\,400} = 1488$

12 Independent for southbound vehicles only.

Under limit and south: $\frac{36}{207} = \frac{54}{207} \times \frac{138}{207}$

Over limit and south: $\frac{18}{207} = \frac{54}{207} \times \frac{69}{207}$

For all 207 vehicles, the ratio 'under limit' to 'over limit' is 2 : 1.

For the southbound vehicles, this ratio is also 2 : 1.

	North	East	South	West	All vehicles
under : over	12 : 5	9 : 5	2 : 1	13 : 7	2 : 1

For vehicles travelling north, east and west, the ratio is not the same as for all 207 vehicles, so the multiplication law does not hold for them.

EXERCISE 4E

1 a Select one letter from BNN: P(select N) = $\frac{2}{3}$

b Select one letter from BAAA: P(select A) = $\frac{3}{4}$

2 a $16 + 48 = 64$ have brothers, and 48 of these have sisters: $\frac{48}{64} = \frac{3}{4}$

b $16 + 12 = 28$ do not have sisters, and 16 of these have brothers: $\frac{16}{28} = \frac{4}{7}$

c $16 + 24 + 12 = 52$ do not have both, and $16 + 24 = 40$ of these have brothers or sisters: $\frac{40}{52} = \frac{10}{13}$

3 a Selection is from 11 colour and 8 black and white photographs: $\frac{12-1}{(12-1)+8} = \frac{11}{19}$

b Selection is from 12 colour and 7 black and white photographs: $\frac{12}{12+(8-1)} = \frac{12}{19}$

4 a i $\frac{4+1}{8+3+4+1} = \frac{5}{16}$ **ii** $\frac{8+4}{8+4+5+6} = \frac{12}{23}$

b Those with an interest in exactly two careers $\left(\frac{5}{9}$ for D; $\frac{6}{9}$ for H; $\frac{7}{9}$ for $N\right)$ or those with an interest in more than one career $\left(\frac{6}{10}$ for D; $\frac{7}{10}$ for H; $\frac{8}{10}$ for $N\right)$.

5 a Select from 39, of whom 20 scored more than 5: $\frac{6+5+5+3+1}{40-1} = \frac{20}{39}$

b Select from 39, of whom 8 scored more than 7: $\frac{5+3+1-1}{40-1} = \frac{8}{39}$

6 a Class frequencies are 10, 40, 45 and 20. $10 + 40 = 50$ men took < 3 min, and 10 of these took < 1 min: $\frac{10}{10+40} = \frac{1}{5}$

b 94 of the 114 we select from took < 6 min: $\frac{(10-1)+40+45}{115-1} = \frac{94}{114} = \frac{47}{57}$ or 0.825

7 a 10% of the staff are part-time and female.

Staff who are not in set M are females, and staff who are not in set FT work part-time.

b $a + b + c = 0.9$ .... [1]
$a + b = 0.6$ .... [2]
$b + c = 0.7$ .... [3]
Substituting [2] into [1] gives $0.6 + c = 0.9$, so $c = 0.3$, $a = 0.2$ and $b = 0.4$.

c i $\frac{0.4}{0.4+0.3} = \frac{4}{7}$ **ii** $\frac{0.3}{0.3+0.1} = \frac{3}{4}$

iii $\frac{0.4}{0.2+0.4+0.3} = \frac{4}{9}$

8 The grid shows sums.

3	4	5	6
2	3	4	5
1	2	3	4
	1	2	3

Five of the sums are even and in three of these the numbers on the spinners are the same: P(same | sum even) = $\frac{3}{5}$

6	7	8	9	10	11	12
5	6	7	8	9	10	11
4	5	6	7	8	9	10
3	4	5	6	7	8	9
2	3	4	5	6	7	8
1	2	3	4	5	6	7
	1	2	3	4	5	6

9 The grid shows scores.
21 scores are greater than 6, and 11 of these are 8 or less: $P(S \leq 8 \mid S > 6) = \frac{11}{21}$

10 a $P(5) = \frac{\pi \times 3^2}{\pi \times 30^2} = \frac{9}{900} = 0.01$ Probabilities are proportional to areas.

b $P(3) = \frac{\pi(9^2 - 3^2)}{\pi \times 30^2} = 0.08$ $P(2) = \frac{\pi(15^2 - 9^2)}{\pi \times 30^2} = 0.16$ $P(1) = \frac{\pi(30^2 - 15^2)}{\pi \times 30^2} = 0.75$

c $P(1 \mid \text{not } 5) = \frac{P(1)}{P(1) + P(2) + P(3)} = \frac{0.75}{0.75 + 0.16 + 0.08} = \frac{25}{33}$ or 0.758

d $P(\text{neither } 1 \mid \text{total } 6) = \frac{P(3, 3)}{P(1, 5) + P(5, 1) + P(3, 3)} = \frac{0.08^2}{(2 \times 0.75 \times 0.01) + 0.08^2} = \frac{32}{107}$ or 0.299

EXERCISE 4F

1 a P(1st plain) × P(2nd plain | 1st plain) = $\frac{3}{8} \times \frac{2}{7} = \frac{3}{28}$

   b P(1st striped) × P(2nd striped | 1st striped) = $\frac{5}{8} \times \frac{4}{7} = \frac{5}{14}$

2 P(toffee and nutty) = P(1st toffee) × P(2nd nutty | 1st toffee) = $\frac{4}{11} \times \frac{7}{10} = \frac{28}{110}$

   P(nutty and toffee) = P(1st nutty) × P(2nd toffee | 1st nutty) = $\frac{7}{11} \times \frac{4}{10} = \frac{28}{110}$

   P(not the same type) = $\frac{28}{110} + \frac{28}{110} = \frac{28}{55}$

3 a P(1st novel) × P(2nd novel | 1st novel) = $\frac{7}{12} \times \frac{6}{11} = \frac{7}{22}$

   b P(1st dict.) × P(2nd dict. | 1st dict.) + P(1st atlas) × P(2nd atlas | 1st atlas)
      = $\frac{3}{12} \times \frac{2}{11} + \frac{2}{12} \times \frac{1}{11} = \frac{2}{33}$

4 a P(bicycle and late) + P(scooter and late) = (0.7 × 0.03) + (0.3 × 0.02) = 0.027
   b P(not late) = 1 − 0.027 = 0.973.
      $n \times 0.973 = 223$ gives $n = 229.188...$, so she works on 229 or 230 days in the year.

5 a P(two boys) = $\frac{5}{12} \times \frac{4}{11} = \frac{20}{132}$ and P(two girls) = $\frac{7}{12} \times \frac{6}{11} = \frac{42}{132}$

Two girls are more likely to be selected because $\frac{42}{132} > \frac{20}{132}$

b For any particular child, P(selected first) = $\frac{1}{12}$ and P(selected second | not selected first) = $\frac{1}{11}$.

P(any two particular children are selected) = $\left(\frac{1}{12} \times \frac{1}{11}\right) + \left(\frac{1}{12} \times \frac{1}{11}\right) = \frac{1}{66}$

The two youngest girls and the two oldest boys are equally likely to be selected.

6 $\frac{1}{10} \times \frac{1}{9} \times 10 = \frac{1}{9}$

7 a P(4, 10) + P(10, 4) + P(7, 7) = $\left(\frac{5}{20} \times \frac{9}{19}\right) + \left(\frac{9}{20} \times \frac{5}{19}\right) + \left(\frac{6}{20} \times \frac{5}{19}\right) = \frac{6}{19}$ or 0.316

b P(7, 7 | total 14) = $\frac{P(\text{total 14 and 7, 7})}{P(\text{total 14})} = \left(\frac{6}{20} \times \frac{5}{19}\right) \div \frac{6}{19} = \frac{1}{4}$

> By rearranging P(A and B) = (P(A) × P(B|A), we can find a conditional probability using
> P(B|A) = $\frac{P(A \text{ and } B)}{P(A)} = \frac{P(B \text{ and } A)}{P(A)}$.

8 a P(win 1st, lose 2nd) + P(lose 1st, lose 2nd) = (0.65 × 0.3) + (0.35 × 0.45)
= 0.3525 or $\frac{141}{400}$

b P(win 1st | lose 2nd) = $\frac{P(\text{lose 2nd and win 1st})}{P(\text{lose 2nd})} = \frac{0.65 \times 0.3}{0.65 \times 0.3 + 0.35 \times 0.45} = 0.553$ or $\frac{26}{47}$

9 a $0.13 \div 0.65 = \frac{1}{5}$ or 0.2

b $0.27 \div 0.81 = \frac{1}{3}$ or 0.333

c $0.35 \div (1 - 0.6) = \frac{7}{8}$ or 0.875

10 P(purchase and bed) = P(purchase) × P(bed | purchase)

P(purchase) = $\frac{P(\text{purchase and bed})}{P(\text{bed | purchase})}$

P(not purchase) = $1 - \frac{P(\text{purchase and bed})}{P(\text{bed | purchase})}$

$= 1 - \frac{0.042}{0.15} = 0.72$

Chapter 4: Probability

11 Of the 91 numbers from 10 to 100 inclusive, 18 contain a 5 and 73 do not.
Of the 73 numbers that do not contain a 5, there are 9 multiples of 5 (10, 20, 30, 40, 60, 70, 80, 90 and 100). Probability is $\frac{9}{73}$ or 0.123.

12 a Let T and X represent a twin and a non-twin, respectively.
$$P(T, X, X) + P(X, T, X) + P(X, X, T) = 3 \times \left(\frac{2}{7} \times \frac{5}{6} \times \frac{4}{5}\right) = \frac{4}{7} \text{ or } 0.571$$

b P(2G and 1B) + P(3G and 0B) = P(G, G, B) + P(G, B, G) + P(B, G, G) + P(G, G, G)
$$= 3 \times \left(\frac{3}{7} \times \frac{2}{6} \times \frac{4}{5}\right) + \left(\frac{3}{7} \times \frac{2}{6} \times \frac{1}{5}\right)$$
$$= \frac{13}{35} \text{ or } 0.371$$

> The three different orders in which two girls and one boy can be selected are equally likely.

13 a $(0.8 \times 0.74) + (1 - 0.8)y = 0.68$
$0.592 + 0.2y = 0.68$, which gives $y = 0.44$.

b P(call not answered) = $(0.8 \times 0.26) + (0.2 \times 0.56)$

$$P(\text{landline} \mid \text{not answered}) = \frac{0.2 \times 0.56}{0.8 \times 0.26 + 0.2 \times 0.56} = 0.35 \text{ or } \frac{7}{20}$$

14 a $0.65x + 0.15(1 - x) = 0.33$ gives $x = 0.36$

b P(offer accepted) = $0.36 \times 0.35 + 0.64 \times 0.85$

$$P(\text{park} \mid \text{accepted}) = \frac{0.64 \times 0.85}{0.36 \times 0.35 + 0.64 \times 0.85} = 0.812 \text{ or } \frac{272}{335}$$

15 Let there be n boys and $(n - 10)$ girls, so there are $(2n - 10)$ children altogether.
There are $(2n - 10)$ possible selections for the 1st child and $(2n - 11)$ for the 2nd.

$(2n - 10)(2n - 11) = 756$
$2n^2 - 21n - 323 = 0$
$(2n + 17)(n - 19) = 0$

$n = 19$, so there are 19 boys and 9 girls.

$$P(\text{2 boys or 2 girls}) = \frac{19}{28} \times \frac{18}{27} + \frac{9}{28} \times \frac{8}{27} = \frac{23}{42} \text{ or } 0.548$$

16 Let W represent 'walks' and let M represent 'meets'.

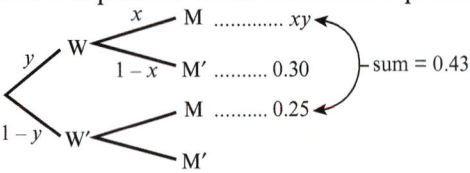

```
        x    M ............ xy
    W <
  y     1−x  M' .......... 0.30   } sum = 0.43
 <           M  .......... 0.25
  1−y   W' <
             M'
```

$xy + 0.25 = 0.43$, so $xy = 0.18$ .... [1]

$y(1 - x) = 0.30$, so $xy = y - 0.30$ .... [2]

Equating [1] and [2] gives $y = P(W) = 0.18 + 0.30 = 0.48$.

17 Aaliyah will not complete the crossword in a magazine if she:
- does not buy a magazine or
- buys a magazine, but does not attempt the crossword or
- buys and attempts, but fails to complete the crossword.

Let A, B and C represent Attempts, Buys and Completes, respectively.

$$P(C') = P(B') + P(B \text{ and } A') + P(B \text{ and } A \text{ and } C') = \frac{2}{7} + \left(\frac{5}{7} \times 0.16\right) + \left(\frac{5}{7} \times 0.84 \times 0.4\right)$$

$$= \frac{16}{25} \text{ or } 0.64$$

Chapter 5
Permutations and combinations

EXERCISE 5A

1 a $\dfrac{5!}{3!} = \dfrac{5 \times 4 \times 3 \times 2 \times 1}{3 \times 2 \times 1} = 5 \times 4 = 20$

b $\dfrac{4!}{2!} - 3! = \dfrac{4 \times 3 \times 2 \times 1}{2 \times 1} - 6 = 12 - 6 = 6$

$\dfrac{4!}{2!} - 3!$ can be factorised: $3!\left(\dfrac{4}{2!} - 1\right) = 6(2-1) = 6$

c $7 \times 4! + 21 \times 3! = (7 \times 3!)(4 + 3) = 7 \times 6 \times 7 = 294$

d $\dfrac{10!}{8!} + \dfrac{9!}{7!} = 10 \times 9 + 9 \times 8 = 90 + 72 = 162$

e $\dfrac{20!}{18!} - \dfrac{13!}{11!} = 20 \times 19 - 13 \times 12 = 380 - 156 = 224$

2 a $9! < 1\,000\,000 < 10!$ so the smallest n is 10.

b $n! > 86\,400$. Now $8! < 86\,400 < 9!$ so the smallest n is 9.

Note: Trial and improvement can be used to answer question **2** and question **3**

c $(3!)! = 6! = 720 < 10^{20}$, but $(4!)! = 24! > 10^{20}$, so the smallest n is 4.

3 a $n! < 40\,000\,000$. Now $11! < 40\,000\,000 < 12!$ so the largest n is 11.

b $n! < 1.5 \times 10^{12}$. Now $15! < 1.5 \times 10^{12} < 16!$ so the largest n is 15.

c $\dfrac{n!}{(n-2)!} = n(n-1)$. Now $22 \times 21 < 500 < 23 \times 22$, so the largest n is 22.

4 $144 = \dfrac{9! \times 2!}{7!} = \dfrac{72! \times 2!}{71!} = \ldots$ $\quad 252 = \dfrac{7! \times 3!}{5!} = \dfrac{126! \times 2!}{125!} = \ldots$ $\quad 1\dfrac{1}{2} = \dfrac{15! \times 4!}{16!} = \dfrac{79! \times 5!}{80!} = \ldots$

5 $53 \times 52 = \dfrac{53!}{51!}$ cm²

6 $(25 \times 24 \times 23) - (8 \times 7 \times 6) = \left(\dfrac{25!}{22!} - \dfrac{8!}{5!}\right)$ cm³

7 $8 \times 7 \times 6 \times 0.09 = \$30.24 = \$\dfrac{9! \times 99!}{5! \times 100!} = \$\dfrac{6! + 3! \times 3!}{4! + 1!} = \$\dfrac{9!}{5! \times (5! - 4! + 2! + 2!)} = \ldots$

EXERCISE 5B

1 Arrange six from six letters:
$^6P_6 = 6! = 6 \times 5 \times 4 \times 3 \times 2 \times 1 = 720$

2 a Arrange 52 from 52 cards: $^{52}P_{52} = 52!$
$= 52 \times 51 \times 50 \times \ldots \ldots \times 2 \times 1 = 8.07 \times 10^{67}$

b Arrange four from four cards:
$^4P_4 = 4! = 4 \times 3 \times 2 \times 1 = 24$

c Arrange 13 from 13 cards:
$^{13}P_{13} = 13! = 13 \times 12 \times 11 \times \ldots \ldots \times 2 \times 1$
$= 6\,227\,020\,800$

3 a Arrange two from two women: $^2P_2 = 2! = 2$
  b Arrange six from six men: $^6P_6 = 6! = 720$
  c Arrange eight from eight people:
     $^8P_8 = 8! = 40\,320$

4 a $^4P_4 = 4! = 24$
  b $^3P_3 = 3! = 6$
  c $^7P_7 = 7! = 5040$

5 They can also be parked in 39 916 800 ways because $7 + x = 5 + x + 2$.

6 The remaining nine children are arranged in the other nine chairs: $^9P_9 = 9! = 362\,880$.

7 $\dfrac{(n+2)!}{n!} = (n+2)(n+1) = 420$

   $n^2 + 3n - 418 = 0$

   $(n + 22)(n - 19) = 0$, so $n = 19$

EXERCISE 5C

1 a Five letters with no repeats: $5! = 120$
  b Six letters with two Ts: $\dfrac{6!}{2!} = 360$
  c Nine letters with two Ms, two Ts and two Es:
     $\dfrac{9!}{2! \times 2! \times 2!} = 45\,360$
  d Eleven letters with four Is, four Ss and two Ps: $\dfrac{11!}{4! \times 4! \times 2!} = 34\,650$
  e Eleven letters with four Ls, two As and two Os: $\dfrac{11!}{4! \times 2! \times 2!} = 415\,800$

2 a Six digits with five 1s: $\dfrac{6!}{5!} = 6$
  b Six digits with three 2s and three 7s:
     $\dfrac{6!}{3! \times 3!} = 20$
  c Six digits with three 6s and two 7s:
     $\dfrac{6!}{3! \times 2!} = 60$
  d Six digits with two 8s and four 9s:
     $\dfrac{6!}{2! \times 4!} = 15$

3 a Three squares with none identical:
     $^3P_3 = 3! = 6$.
  b Five squares with five identical in shape and colour: $\dfrac{5!}{5!} = 1$

  c Fifteen squares with seven identical blue and eight identical green: $\dfrac{15!}{7! \times 8!} = 6435$
  d Twenty squares with five identical red, seven identical blue and eight identical green:
     $\dfrac{20!}{5! \times 7! \times 8!} = 99\,768\,240$

4 The first student is correct. The second student has treated the trees as two identical objects and the bushes as three identical objects.

> If, for example, we consider colours only then a red object and two green objects can be arranged in a row in three ways: RGG, GRG and GGR.
> However, if we consider the actual objects, then a red tomato, a green apple and a green grape can be arranged in a row in six ways: TAG, TGA, ATG, AGT, GTA and GAT.

5 a Two possibilities for each of the ten coins:
     $2^{10} = 1024$
  b i Ten coins with five heads and five tails:
        $\dfrac{10!}{5! \times 5!} = 252$
     ii Of the 1024 arrangements, 252 show the same number of heads as tails, so half the remaining arrangements show more heads than tails: $\dfrac{1024 - 252}{2} = 386$.

6 $420 = \frac{7!}{12} = \frac{7!}{3! \times 2!}$, so one letter appears three times, another letter appears twice, and the other two letters appear once each, e.g. peepers.

7 a Five letters with two As and three Bs:
$\frac{5!}{2! \times 3!} = 10$

b There are five vowels to choose from:
$5 \times \frac{5!}{2! \times 3!} = 50$

c There are five vowels and 21 consonants to choose from: $5 \times 21 \times \frac{5!}{2! \times 3!} = 1050$

EXERCISE 5D

1 a $^5P_5 = 5! = 120$

   b i The number must end in 3 or 5 (two choices), and the remaining four digits can be arranged in $^4P_4$ ways:
   $^4P_4 \times 2 = 4 \times 3 \times 2 \times 1 \times 2 = 4! \times 2 = 48$.

   ii The number must end in 2, 4 or 6 (three choices) and the remaining four digits can be arranged in $^4P_4$ ways:
   $^4P_4 \times 3 = 4 \times 3 \times 2 \times 1 \times 3 = 4! \times 3 = 72$.

   Alternatively, we can subtract the number of odd numbers from $^5P_5$, which gives $120 - 48 = 72$.

   iii The numbers must end in 3 or 5 and must begin with 2 or 3 (note that 3 can be at the end or at the beginning).
   After placing first and last digits, the other three digits can be arranged in $^3P_3$ ways.
   Ends in 3 (begins with 2): $1 \times {}^3P_3 \times 1 = 6$
   Ends in 5 (begins with 2 or 3): $2 \times {}^3P_3 \times 1 = 12$
   Total is $6 + 12 = 18$.

2 a The two women can be at the front in $^2P_2$ ways, and the four men can be arranged behind them in $^4P_4$ ways. Total is $^2P_2 \times {}^4P_4 = 48$.

   b A woman can be placed at the front in $^2P_1$ ways, and a man can be placed at the back in $^4P_1$ ways. The other four people can be arranged between them in $^4P_4$ ways.
   Total is $^2P_1 \times {}^4P_4 \times {}^4P_1 = 192$.

   c We can find the number of arrangements in which the two women are not separated then subtract this from the total number of possible arrangements of six people.
   Total number of possible arrangements of six people is $^6P_6$.
   The two women can be next to each other in $^2P_2$ ways.
   The block of two women can be arranged with the four men in $^5P_5$ ways.
   Total is $^6P_6 - ({}^2P_2 \times {}^5P_5) = 480$.

   d The four men can be arranged in $^4P_4$ ways. This block of four men can be arranged with the two women in $^3P_3$ ways.
   Total is $^4P_4 \times {}^3P_3 = 144$.

   e None. There are only three spaces between or on either side of the two women, so the four men cannot all be separated.

3 All possible arrangements are equally likely, so there are an equal number of arrangements ending in each of the six digits.
   Four of the digits are odd and two are even, so the ratio is $4 : 2 = 2 : 1$.

4   a   The two oldest books can be arranged in the middle in $^2P_2$ ways, and the other eight books can be arranged in $^8P_8$ ways.
Total is $^2P_2 \times {}^8P_8 = 80\,640$.

    b   The three newest books can be arranged in $^3P_3$ ways, and this block of three books can be arranged with the other seven books in $^8P_8$ ways.
Total is $^3P_3 \times {}^8P_8 = 241\,920$.

5   a   We can subtract the number of arrangements where the calves are in adjacent stalls from the total number of possible arrangements of the seven animals.
Arrange seven animals: $^7P_7$
Arrange the calves in adjacent stalls: $^2P_2$
Arrange this pair of calves with the 5 cows: $^6P_6$
Total is $^7P_7 - ({}^2P_2 \times {}^6P_6) = 3600$.

    b   Arrange the calves and their mother in a row: $^3P_3$
Arrange this family group with the remaining 4 cows: $^5P_5$
Total is $^3P_3 \times {}^5P_5 = 720$.

    c   Arrange the two calves so that both are next to their mother: $^2P_2$
Arrange this family group with the remaining 4 cows: $^5P_5$
Total is $^2P_2 \times {}^5P_5 = 240$.

6   a   There are $\dfrac{6!}{2! \times 3!}$ distinct six-digit numbers.
Two of the six digits are 2s, so $\dfrac{2}{6}$ of the six-digit numbers begin with a 2.
Total is $\dfrac{2}{6} \times \dfrac{6!}{2! \times 3!} = 20$.

> Alternatively, place a 2 at the left then arrange the other five digits: $1 \times \dfrac{5!}{3!} = 20$.

    b   Numbers not divisible by 2 are odd, so they must end in 1 or 3.

Four of the six digits are 1 or 3, so $\dfrac{4}{6}$ of the numbers are not divisible by 2.
Total is $\dfrac{4}{6} \times \dfrac{6!}{2! \times 3!} = 40$.

> Alternatively,
> Ends in 1: $\dfrac{5!}{2! \times 3!} \times 1 = 10$
> Ends in 3: $\dfrac{5!}{2! \times 2!} \times 1 = 30$
> Total is $10 + 30 = 40$.

7   a   Place two Ts at the beginning and two Es at the end.
Arrange the three remaining letters (H, A and R) between the Ts and Es: $^3P_3$.
Total is $1 \times {}^3P_3 \times 1 = 6$.

    b   Arrange T, E, A, T, R, E in a row then place H in the middle: $\dfrac{6!}{2! \times 2!} \times 1 = 180$.

    c   Arrange T, H, T, R at the beginning and arrange E, A, E at the end.
Total is $\dfrac{4!}{2!} \times \dfrac{3!}{2!} = 36$.

8   a   One choice for the placement of each parcel: $1 \times 1 \times 1 \times 1 \times 1 = 1$.

    b   If only one parcel is in the wrong box then four parcels are in the correct boxes. If, for example, A, B, C and D are placed correctly, that leaves parcel E to be placed in the wrong box, but the only box available is E, which is the correct box! Therefore, it is not possible to place exactly one parcel in the wrong box. The answer is 0.

    c   Mr A's parcel and one other person's parcel placed correctly: four ways.
All three remaining parcels can be placed in the wrong boxes in two ways.
Total is $4 \times 2 = 8$.

    d   Two of the five parcels can be placed in the correct boxes in ten ways (AB, AC, AD, AE, BC, BD, BE, CD, CE or DE). All of the remaining three parcels can be placed in the wrong boxes in two ways.
Total is $10 \times 2 = 20$.

Chapter 5: Permutations and combinations

> Alternatively, $\dfrac{\text{answer c} \times 5}{2!} = \dfrac{8 \times 5}{2!} = 20$.

9 y girls have $(y+1)$ spaces between or on either side of them.

If there are more than $(y+1)$ boys then they cannot all be separated.

It is not possible to separate all of the boys if $x > y + 1$ or $x \geq y + 2$ or equivalent.

EXERCISE 5E

1 a $^7P_5 = 2520$
   b $^9P_4 = 3024$

2 Select and arrange six from 12 books: $^{12}P_6 = 665\,280$.

3 Select three from 20 athletes and arrange/award the three medals: $^{20}P_3 = 6840$.

4 a Select two from 14 colours and arrange them on the two doors: $^{14}P_2 = 182$.

> Alternatively, 14 choices for the first door and 13 choices for the second: $14 \times 13 = 182$.

   b There are 14 choices for each door: $14^2 = 196$.

5 a All of the $^6P_4$ possible arrangements are equally likely, and equal numbers of them begin with each of the six letters.
   One-sixth of the arrangements will begin with A: $\dfrac{1}{6} \times {}^6P_4 = 60$.

> Alternatively, place the A at the beginning then select and arrange three of the remaining five letters after it: $1 \times {}^5P_3 = 60$.

   b The A can appear as the 1st, 2nd, 3rd or 4th letter in the arrangement: $\dfrac{4}{6} \times {}^6P_4 = 240$.

> Alternatively, place the A in any of the four positions then select and arrange three of the remaining five letters: $^4P_1 \times {}^5P_3 = 240$.

6 a Select two from 17 children and arrange as hero and villain: $^{17}P_2 = 272$.
   b Select and arrange two from seven girls or two from ten boys: $^7P_2 + {}^{10}P_2 = 132$.
   c $272 - 132 = 140$

7 a Select and arrange six from nine rings: $^9P_6 = 60\,480$.
   b Arrange the two least expensive rings: $^2P_2$.
   Select and arrange four of the remaining seven rings: $^7P_4$.
   Total is $^2P_2 \times {}^7P_4 = 1680$.

8 One of the three even digits must go at the right: $^3P_1$.
   Select and arrange three of the remaining six digits: $^6P_3$.
   Total is $^6P_3 \times {}^3P_1 = 360$.

9 a Number must end in 0: $^1P_1$.
   Select and arrange two of the remaining four digits: $^4P_2$.
   Total is $^4P_2 \times {}^1P_1 = 12$.
   b Place one of the four non-zero digits at the left, then select and arrange two of the remaining four digits. Total is $^4P_1 \times {}^4P_2 = 48$.

10 E.g. 120 ways for 1st, 2nd and 3rd places to be decided in a race between six athletes or 120 ways to arrange a hand of five cards. [Can also use $^5P_4$ or $^{120}P_1$].

11 a $^nP_r > {^nP_{n-r}}$ means $\dfrac{n!}{(n-r)!} > \dfrac{n!}{r!}$

$$n!\,r! > n!(n-r)!$$
$$r! > (n-r)!$$
$$r > n-r,\text{ so } r > \tfrac{1}{2}n$$

b $^nP_r \times {^nP_{n-r}} = k \times {^nP_n}$ means $\dfrac{n!}{(n-r)!} \times \dfrac{n!}{[n-(n-r)]!} = k \times \dfrac{n!}{0!}$

$$\dfrac{n!}{(n-r)!} \times \dfrac{n!}{r!} = k \times n!$$

$$k = \dfrac{n!}{r!(n-r)!}$$

12 Select and arrange three from 52 cards (selection of five is irrelevant): $^{52}P_3 = 132\,600$

13 Chairs B, D and F can be occupied in $^3P_3$ ways.

Arrange four of the remaining nine people in chairs A, C, E and G: $^9P_4$.

Total is $^3P_3 \times {^9P_4} = 18\,144$.

14 a $^{11}P_8 = 6\,652\,800$

b Place the particular passenger in a shady seat: $^5P_1$.

Arrange the remaining seven passengers in the other ten seats: $^{10}P_7$.

Total is $^5P_1 \times {^{10}P_7} = 3\,024\,000$.

c Let the two particular passengers be X and Y.

For each seat that X can occupy, we can find the number of seats that Y can sit in.

shady | 8 | 7 | 7 | 7 | 8 |
sunny | 9 | 7 | 7 | 7 | 7 | 8 |

↑ If X sits here, there are 9 seats that Y can sit in.

There are $9 + (3 \times 8) + (7 \times 7) = 82$ possible arrangements of X and Y.

The remaining six people can be arranged in the other nine seats in $^9P_6$ ways.

Total is $82 \times {^9P_6} = 4\,959\,360$.

EXERCISE 5F

1 a Select or choose five apples from eight apples: $^8C_5 = 56$.

b Select or choose five apples from nine apples: $^9C_5 = 126$.

2 a Select four from seven men, and five from eight women: $^7C_4 \times {^8C_5} = 1960$.

b Select three from seven men, and six from eight women: $^7C_3 \times {^8C_6} = 980$.

c Select 13 or 14 or 15 people from 15 people: $^{15}C_{13} + {^{15}C_{14}} + {^{15}C_{15}} = 121$.

Chapter 5: Permutations and combinations

3 a Select five from 52 cards: $^{52}C_5 = 2\,598\,960$.
   b Select three from 26 red cards, and two from 26 black cards: $^{26}C_3 \times {}^{26}C_2 = 845\,000$.

4 a i Select six from 26: $^{26}C_6 = 230\,230$ (or select 20 to ignore from 26).
      ii Select 20 from 26: $^{26}C_{20} = 230\,230$ (or select six to ignore from 26).
   b $x = y + z$

5 Two positions for each of the four lights: $2^4 = 16$.

6 Favourable selections are none from six boys and three from seven girls, or one from six boys and two from seven girls:

	Boys (from 6)	Girls (from 7)	
	0	3	$^6C_0 \times {}^7C_3 = 35$
or	1	2	$^6C_1 \times {}^7C_2 = 126$
			Total = 161

7 a Select one from six red, one from five blue, and one from four yellow:
      Total is $^6C_1 \times {}^5C_1 \times {}^4C_1 = 120$.
   b Select three from six red or three from five blue or three from four yellow.
      Total is $^6C_3 + {}^5C_3 + {}^4C_3 = 34$.
   c Select ten from the 11 red and blue or ten from the ten red and yellow.
      Total is $^{11}C_{10} + {}^{10}C_{10} = 12$.
   d Select nine from the 11 red and blue or nine from the ten red and yellow or nine from the nine blue and yellow. Total is $^{11}C_9 + {}^{10}C_9 + {}^9C_9 = 66$.

8 Choices: Morning (3); afternoon (4 + 1 = 5); evening (3). Total is $3 \times 5 \times 3 = 45$.

9 They can share the taxis in 56 ways, no matter which is occupied first.

10 a Select ten of the 20 spaces to leave empty: $^{20}C_{10} = 184\,756$.
   b 1st row empty and 2nd row full or 1st row full and 2nd row empty: two arrangements.
   c Select five spaces to leave empty in each row: $^{10}C_5 \times {}^{10}C_5 = 63\,504$.
   d Select six spaces to leave empty in the 1st row, and four spaces to leave empty in the 2nd row, or vice versa.
      Total is $({}^{10}C_6 \times {}^{10}C_4) + ({}^{10}C_4 \times {}^{10}C_6) = 88\,200$.

11 Without restrictions, he has $8 \times 7 \times 6 = 336$ choices for dressing.
   The red trousers can be worn with the red shirt in six ways (with each jacket).
   There are $336 - 6 = 330$ ways in which he can dress.

12 Select six from ten objects and arrange them with the clock: $^{10}C_6 \times {}^7P_7 = 1\,058\,400$.

   Alternatively, select and arrange six of the ten objects, then place the clock in one of the seven positions between or on either side of them:
   $^{10}P_6 \times {}^7P_1 = 1\,058\,400$.

13 a Select five from ten posters: $^{10}C_5 = 252$.
   b Select five from the remaining eight posters: $^8C_5 = 56$.
   c

	Posters			
	A	C	T	
	1	1	3	$^3C_1 \times {}^2C_1 \times {}^5C_3 = 60$
or	1	2	2	$^3C_1 \times {}^2C_2 \times {}^5C_2 = 30$
or	2	2	1	$^3C_2 \times {}^2C_2 \times {}^5C_1 = 15$
or	2	1	2	$^3C_2 \times {}^2C_1 \times {}^5C_2 = 60$
or	3	1	1	$^3C_3 \times {}^2C_1 \times {}^5C_1 = 10$
				Total = 175

By selecting one of each type then two from the remaining seven, we obtain $^3C_1 \times {}^2C_1 \times {}^5C_1 \times {}^7C_2 = 630$ possible selections. However, this is more than the 252 possible selections that can be made with no restrictions, so this method is clearly not valid, as many of the combinations are counted more than once.

14

Encryption	E	C	H	K	B	P	U	J	S	N	O	L
Letter	S	A	T	U	R	N	↓	↓	↓	↓	↓	↓
No. choices	1	1	1	1	1	1	20	19	18	17	16	15

We have six letters to find from the 20 letters that remain: $^{20}P_6 = 27\,907\,200$.

15 No 2s: arrange three from 1, 3, 4, 5: $^4P_3 = 24$.

One 2: select two from 1, 3, 4, 5 and arrange with the 2: $^4C_2 \times 3! = 36$.

Two 2s: select one from 1, 3, 4, 5 and arrange with the two 2s: $^4C_1 \times \frac{3!}{2!} = 12$.

Total is $24 + 36 + 12 = 72$.

16 a $^3C_2 \times {^6C_1} = 18$

 b There are $^3C_1$ ways to select one set of twins.

 For the other three people, there are three possibilities:

 3 girls: $^4C_3$

 or 1 twin and 2 girls: $^4C_1 \times {^4C_2}$

 or 2 who are twins (but not each other's twin) and 1 girl: $4 \times {^4C_1}$.

 Total is $^3C_1 \times [^4C_3 + (^4C_1 \times {^4C_2}) + (4 \times {^4C_1})] = 132$.

EXERCISE 5G

1 a Total number of possible selections (two from ten children): $^{10}C_2$.

Favourable selections (two from six boys and none from four girls): $^6C_2 \times {^4C_0}$.

$\frac{^6C_2 \times {^4C_0}}{^{10}C_2} = \frac{1}{3}$

 b Favourable selections are none from six boys and two from four girls: $^6C_0 \times {^4C_2}$.

$\frac{^6C_0 \times {^4C_2}}{^{10}C_2} = \frac{2}{15}$

 c Favourable selections are one from six boys and one from four girls: $^6C_1 \times {^4C_1}$.

$\frac{^6C_1 \times {^4C_1}}{^{10}C_2} = \frac{8}{15}$

2 a Total number of possible selections (three from 25 chocolates): $^{25}C_3$.

Favourable selections are one from ten milk and two from 15 dark: $^{10}C_1 \times {^{15}C_2}$.

$\frac{^{10}C_1 \times {^{15}C_2}}{^{25}C_3} = \frac{21}{46}$ or 0.457

 b Favourable selections are two from ten milk and one from 15 dark: $^{10}C_2 \times {^{15}C_1}$.

$\frac{^{10}C_2 \times {^{15}C_1}}{^{25}C_3} = \frac{27}{92}$ or 0.293

 c Favourable selections are one from ten milk and two from 15 dark or two from ten milk and one from 15 dark: $(^{10}C_1 \times {^{15}C_2}) + (^{10}C_2 \times {^{15}C_1})$.

$\frac{^{10}C_1 \times {^{15}C_2}}{^{25}C_3} + \frac{^{10}C_2 \times {^{15}C_1}}{^{25}C_3} = \frac{21}{46} + \frac{27}{92} = \frac{3}{4}$

3 a Total number of possible selections: $^{40}C_4$.

Favourable selections are four from 17 yellow and none from 23 green: $^{17}C_4 \times {^{23}C_0}$.

$\frac{^{17}C_4 \times {^{23}C_0}}{^{40}C_4} = 0.0260$

 b Favourable selections are four from 17 yellow and none from 23 green, or three from 17 yellow and one from 23 green: $(^{17}C_4 \times {^{23}C_0}) + (^{17}C_3 \times {^{23}C_1})$.

$\frac{^{17}C_4 \times {^{23}C_0}}{^{40}C_4} + \frac{^{17}C_3 \times {^{23}C_1}}{^{40}C_4} = 0.197$

4 Total number of possible selections: $^{80}C_8$.

Favourable selections from 36P and 44S are: 8P and 0S, 7P and 1S, or 6P and 2S.

$$\frac{^{36}C_8 \times ^{44}C_0}{^{80}C_8} + \frac{^{36}C_7 \times ^{44}C_1}{^{80}C_8} + \frac{^{36}C_6 \times ^{44}C_2}{^{80}C_8} = 0.0773$$

5 Total number of possible selections is $^{100}C_5$.

Favourable selections from 67W and 33M are: 5W and 0M or 3W and 2M or 1W and 4M.

$$\frac{^{67}C_5 \times ^{33}C_0}{^{100}C_5} + \frac{^{67}C_3 \times ^{33}C_2}{^{100}C_5} + \frac{^{67}C_1 \times ^{33}C_4}{^{100}C_5} = 0.501$$

6 Total number of possible selections is $^{90}C_4$.

Favourable selections are none from 11 chisels and four from 79 non-chisels.

$$\frac{^{11}C_0 \times ^{79}C_4}{^{90}C_4} = 0.588$$

7 a All combinations of red and blue wigs are equally likely with probability $\left(\frac{1}{2}\right)^5 = \frac{1}{32}$.

$^5C_2 \times \frac{1}{32} = \frac{5}{16}$

b $(^5C_5 + ^5C_4 + ^5C_3) \times \frac{1}{32} = \frac{1}{2}$

8 a The nine rose bushes are equally likely to be in the middle: $\frac{6}{9} = \frac{2}{3}$

b Total number of arrangements: $^9P_9$.

Arrangements of three red in a row: $^3P_3$.

Arrangements of the block of three red with the six yellow: $^7P_7$.

$$\frac{^3P_3 \times ^7P_7}{^9P_9} = \frac{1}{12}$$

c Arrange six yellow in a row: $^6P_6$.

Select three of the seven spaces between or on either side of the yellow bushes: $^7P_3$.

$$\frac{^6P_6 \times ^7P_3}{^9P_9} = \frac{5}{12}$$

9 a Total number of possible selections is $^{50}C_6$.

Favourable selections are three from 24 sheep and three from 26 cattle.

$$\frac{^{24}C_3 \times ^{26}C_3}{^{50}C_6} = 0.331$$

b Favourable selections from 41F and 9M are 6F and 0M or 5F and 1M or 4F and 2M.

$$\frac{^{41}C_6 \times ^9C_0}{^{50}C_6} + \frac{^{41}C_5 \times ^9C_1}{^{50}C_6} + \frac{^{41}C_4 \times ^9C_2}{^{50}C_6} = 0.937$$

10 a Ten letters with three Ss, three Ts and two Is:

$$\frac{10!}{3! \times 3! \times 2!} = 50\,400.$$

b i Arrange the letters SAISICS after the three Ts: $\frac{7!}{3! \times 2!}$.

Probability is $\frac{7!}{3! \times 2!} \div 50\,400 = \frac{1}{120}$.

ii Begins TTT: arrange the letters SAISICS

$\frac{7!}{3! \times 2!}$.

Begins SSS: arrange the letters TATITIC

$\frac{7!}{3! \times 2!}$.

Probability is

$$\left(\frac{7!}{3! \times 2!} + \frac{7!}{3! \times 2!}\right) \div 50\,400 = \frac{1}{60}.$$

11 a Total number of possible arrangements: $^9P_9$.

Arrange three skirts in the middle: $^3P_3$.

Arrange the other six items: $^6P_6$.

Probability is $\frac{^3P_3 \times ^6P_6}{^9P_9} = \frac{1}{84}$ or 0.0119.

b Arrange two jackets next to each other: $^2P_2$.

Arrange the block of two jackets with the other seven items: $^8P_8$.

$$\frac{^2P_2 \times ^8P_8}{^9P_9} = \frac{2}{9}$$

12 Total number of possible selections: $^{180}C_6$.

There are 101 people who are either left handed or female and 79 who are not.

Favourable selections are four from 101 and two from 79.

Probability is $\frac{^{101}C_4 \times ^{79}C_2}{^{180}C_6} = 0.290$

13 a $a = 478 - 312 = 166$

$a + b = 440$, so $b = 274$

$b + c = 1240 - 478$, so $c = 488$.

b There are 966 books that are novels or hard covers (or both) and 274 that are not.

$$\frac{(^{966}C_{22} \times {}^{274}C_3) + (^{966}C_{23} \times {}^{274}C_2) + (^{966}C_{24} \times {}^{274}C_1) + (^{966}C_{25} \times {}^{274}C_0)}{^{1240}C_{25}} = 0.162$$

14 Total number of possible selections: $(^{10}C_5 \times {}^5C_2) + (^{10}C_6 \times {}^5C_1) + (^{10}C_7 \times {}^5C_0)$.

Favourable selections are 5W and 2M: $^{10}C_5 \times {}^5C_2$.

$$\frac{^{10}C_5 \times {}^5C_2}{(^{10}C_5 \times {}^5C_2) + (^{10}C_6 \times {}^5C_1) + (^{10}C_7 \times {}^5C_0)} = \frac{2520}{3690} = \frac{28}{41} \text{ or } 0.683$$

15 Let there be t tags and l labels in the box.

$$\frac{^tC_2}{^{(t+l)}C_2} = 5 \times \frac{^lC_2}{^{(t+l)}C_2} \text{ gives } t(t-1) = 5l^2 - 5l \ldots\ldots\ldots\ldots \text{ [1]}$$

$$\frac{^tC_1 \times {}^lC_1}{^{(t+l)}C_2} = 6 \times \frac{^lC_2}{^{(t+l)}C_2} \text{ gives } t = 3l - 3 \ldots\ldots\ldots\ldots\ldots \text{ [2]}$$

Substituting [2] into [1] gives $(3l-3)(3l-4) = 5l^2 - 5l$

$$l^2 - 4l + 3 = 0$$

$$(l-1)(l-3) = 0, \text{ so } l = 1 \text{ or } l = 3$$

$l = 1$ gives $t = 0$, which is invalid, and $l = 3$ gives $t = 6$.

The box contains three labels and six tags.

16 a For $n = 2$, $P(X) = \frac{1 \times 2!}{3!} = \frac{1}{3}$ and $P(X') = \frac{2}{3}$

For $n = 3$, $P(X) = \frac{2 \times 3!}{4!} = \frac{2}{4}$ and $P(X') = \frac{2}{4}$

For $n = 4$, $P(X) = \frac{3 \times 4!}{5!} = \frac{3}{5}$ and $P(X') = \frac{2}{5}$

For $n = 5$, $P(X) = \frac{4 \times 5!}{6!} = \frac{4}{6}$ and $P(X') = \frac{2}{6}$

$P(X) = \frac{n-1}{n+1}$ and $P(X') = 1 - P(X) = \frac{2}{n+1}$

b $\frac{P(X')}{P(X)} = \frac{2}{n+1} \div \frac{n-1}{n+1} = \frac{2}{n-1}$

There are $(n+1)!$ arrangements of n pizzas and one pasta dish. The pasta dish is not between two pizzas when it is at the left side or at the right side of the n pizzas, which themselves can be arranged in $n!$ ways. So, there are $2 \times n!$ arrangements favourable to X', and $P(X') = \frac{2 \times n!}{(n+1)!}$.

There are $(n+1)! - 2 \times n!$ arrangements favourable to X, so $P(X) = \frac{(n+1)! - 2 \times n!}{(n+1)!}$.

$$\frac{P(X')}{P(X)} = \left(\frac{2 \times n!}{(n+1)!}\right) \div \left(\frac{(n+1)! - 2 \times n!}{(n+1)!}\right) = \frac{2 \times n!}{(n+1)! - 2 \times n!} = \frac{2 \times n!}{n![n+1-2]} = \frac{2}{n-1} \text{ for any } n \geqslant 2.$$

Chapter 6
Probability distributions

EXERCISE 6A

1 Let $P(V=3) = p$ then $P(V=2) = 2p$ and $P(V=1) = 2p$.

Sum of probabilities is $2p + 2p + p = 1$, which gives $p = 0.2$.

v	1	2	3
P(V=v)	0.4	0.4	0.2

2 Sum of probabilities is $p + 2p + \frac{1}{2}p + 3p = 1$, which gives $p = \frac{2}{13}$.

x	2	3	4	5
P(X=x)	$\frac{2}{13}$	$\frac{4}{13}$	$\frac{1}{13}$	$\frac{6}{13}$

$P(2 < X < 5) = P(X = 3 \text{ or } 4) = \frac{4}{13} + \frac{1}{13} = \frac{5}{13}$

3 a Sum of probabilities is $2k + k^2 + \frac{k}{2} + \left(\frac{4}{5} - 3k\right) + \frac{13}{50} = 1$ which gives:

$50k^2 - 25k + 3 = 0$
$(10k - 3)(5k - 1) = 0$, so $k = 0.2$ or $k = 0.3$

b Probability distributions using $k = 0.2$ and $k = 0.3$ are:

w		3	6	9	12	15	
P(W=w)	k=0.2	0.4	0.04	0.1	0.2	0.26	valid
	k=0.3	0.6	0.09	0.15	−0.1	0.26	invalid

Probabilities cannot be negative, so $k = 0.2$ is the only valid solution.

c $P(6 \leq W < 10) = P(W = 6 \text{ or } 9) = 0.04 + 0.1 = 0.14$

4 $P(S=0) = \frac{2}{9} \times \frac{2}{9} = \frac{4}{81}$ $P(S=1) = \frac{7}{9} \times \frac{2}{9} + \frac{2}{9} \times \frac{7}{9} = \frac{28}{81}$ $P(S=2) = \frac{7}{9} \times \frac{7}{9} = \frac{49}{81}$

s	0	1	2
P(S=s)	$\frac{4}{81}$	$\frac{28}{81}$	$\frac{49}{81}$

5 a Let R be the number of red roses, then $P(R=3) = \dfrac{{}^{25}C_3 \times {}^{40}C_0}{{}^{65}C_3} = 0.0527$ (3 significant figures).

Alternatively,
$P(R=3) = \frac{25}{65} \times \frac{24}{64} \times \frac{23}{63} = 0.0527$.

b $P(R=0) = \dfrac{^{25}C_0 \times {}^{40}C_3}{^{65}C_3} = 0.226$ $P(R=1) = \dfrac{^{25}C_1 \times {}^{40}C_2}{^{65}C_3} = 0.446$ $P(R=2) = \dfrac{^{25}C_2 \times {}^{40}C_1}{^{65}C_3} = 0.275$

r	0	1	2	3
P(R = r)	0.226	0.446	0.275	0.0527

c $P(R \geq 1) = 1 - P(R=0) = 1 - 0.22619\ldots = 0.774$

6 a $P(\text{three vans}) = P(V=3) = \dfrac{^5C_3 \times {}^{10}C_0}{^{15}C_3} = \dfrac{10}{455} = \dfrac{2}{91}$

b $P(V=0) = \dfrac{^5C_0 \times {}^{10}C_3}{^{15}C_3} = \dfrac{24}{91}$ $P(V=1) = \dfrac{^5C_1 \times {}^{10}C_2}{^{15}C_3} = \dfrac{45}{91}$ $P(V=2) = \dfrac{^5C_2 \times {}^{10}C_1}{^{15}C_3} = \dfrac{20}{91}$

v	0	1	2	3
P(V = v)	$\dfrac{24}{91}$	$\dfrac{45}{91}$	$\dfrac{20}{91}$	$\dfrac{2}{91}$

c $P(V \leq 1) = P(V=0 \text{ or } 1) = \dfrac{24}{91} + \dfrac{45}{91} = \dfrac{69}{91}$

7 R, the number of red grapes selected: $R \in \{0, 1\}$
G, the number of green grapes selected: $G \in \{4, 5\}$ $\Big\} R + G = 5$

8 For each selection of three DVDs, $M + D = 3$, so $P(D = d) = P(M = 3 - d)$.

d	0	1	2
P(D = d)	0.1	0.6	0.3

9 $P(\text{right-handed and has red hair}) = 0.9 \times 0.4 = 0.36$
$P(X=0) = 0.64 \times 0.64 = 0.4096$
$P(X=1) = (0.64 \times 0.36) + (0.36 \times 0.64) = 0.4608$
$P(X=2) = 0.36 \times 0.36 = 0.1296$

x	0	1	2
P(X = x)	0.4096	0.4608	0.1296

We assume that being right-handed and having red hair are independent.

10 a $P(X=8) = P(3 \text{ and } 5) + P(5 \text{ and } 3) = \dfrac{1}{4} \times \dfrac{1}{4} + \dfrac{1}{4} \times \dfrac{1}{4} = \dfrac{1}{8}$

b The grid shows the possible sums.

5	6	7	8	10
3	4	5	6	8
2	3	4	5	7
1	2	3	4	6
	1	2	3	5

x	2	3	4	5	6	7	8	10
P(X = x)	$\frac{1}{16}$	$\frac{2}{16}$	$\frac{3}{16}$	$\frac{2}{16}$	$\frac{3}{16}$	$\frac{2}{16}$	$\frac{2}{16}$	$\frac{1}{16}$

$P(X > 6) = P(X = 7, 8 \text{ or } 10) = \frac{2}{16} + \frac{2}{16} + \frac{1}{16} = \frac{5}{16}$

11 a Only three letters are not addressed to Mr Nut, so the probability is 0.

b $P(N = 1) = \frac{{}^5C_1 \times {}^3C_3}{{}^8C_4} = \frac{1}{14}$ $P(N = 2) = \frac{{}^5C_2 \times {}^3C_2}{{}^8C_4} = \frac{6}{14}$

$P(N = 3) = \frac{{}^5C_3 \times {}^3C_1}{{}^8C_4} = \frac{6}{14}$ $P(N = 4) = \frac{{}^5C_4 \times {}^3C_0}{{}^8C_4} = \frac{1}{14}$

n	1	2	3	4
P(N = n)	$\frac{1}{14}$	$\frac{6}{14}$	$\frac{6}{14}$	$\frac{1}{14}$

c Both diagrams would be symmetric.

12 $P(Y = 8, 9 \text{ or } 10) = 8k + 9k + 10k = 1$, which gives $k = \frac{1}{27}$.

13 a $P(Q = 3, 4, 5 \text{ or } 6) = 9c + 16c + 25c + 36c = 1$, which gives $c = \frac{1}{86}$.

b $P(Q > 4) = P(Q = 5 \text{ or } 6) = \frac{25}{86} + \frac{36}{86} = \frac{61}{86}$

14 a $P(N = 2) = \frac{{}^{10}C_2 \times {}^{15}C_2}{{}^{25}C_4} = 0.374$

b $N = 0$ is more likely than $N = 4$. Each time a book is selected, $P(N') > P(N)$ or there are always fewer novels than non-novels in the box.

15 a $P(X = 0) = P(0 \text{ on first and } 0 \text{ on second}) = \frac{1}{4} \times \frac{1}{3} = \frac{1}{12}$

b $P(X = 1) = \frac{1}{4} + \left(\frac{1}{4} \times \frac{1}{3}\right) = \frac{1}{3}$ $P(X = 2) = \frac{1}{4} + \left(\frac{1}{4} \times \frac{1}{3}\right) = \frac{1}{3}$ $P(X = 3) = \frac{1}{4}$

x	0	1	2	3
P(X = x)	$\frac{1}{12}$	$\frac{4}{12}$	$\frac{4}{12}$	$\frac{3}{12}$

$P(X \text{ is prime}) = P(X = 2 \text{ or } 3) = \frac{7}{12}$

16 a $\sqrt[3]{a} = \sqrt[3]{1 - (0.512 + 0.384 + 0.096)} = 0.2$

b The obvious discrete random variable is T, the number of tails obtained (but others, such as $2H$, $0.5H$ or $10-3H$, etc. are valid).

h	0	1	2	3
t	3	2	1	0
P(H = h) and P(T = t)	0.512	0.384	0.096	0.008

$P(T > H) = P(H = 0 \text{ or } 1) = P(T = 3 \text{ or } 2) = 0.512 + 0.384 = 0.896$

17 a The grid shows points awarded.

6	1	2	1	2	3	2
5	1	1	1	3	1	3
4	1	2	3	2	3	2
3	1	3	1	3	1	1
2	3	2	3	2	1	2
1	1	3	1	1	1	1
	1	2	3	4	5	6

s	1	2	3
P(S = s)	$\frac{17}{36}$	$\frac{9}{36}$	$\frac{10}{36}$

b There are six ways to obtain a sum > 9, and two of these score 3 points.

$P(3 \text{ points} \mid \text{sum} > 9) = \frac{2}{6} = \frac{1}{3}$

Alternatively,

$P(3 \text{ points} \mid \text{sum} > 9) = \dfrac{P(\text{sum} > 9 \text{ and } 3 \text{ points})}{P(\text{sum} > 9)} = \dfrac{2}{36} \div \dfrac{6}{36} = \dfrac{1}{3}$

18 a $P(R = 1, 3, 5 \text{ or } 7) = \dfrac{2k}{3} + \dfrac{4k}{5} + \dfrac{6k}{7} + \dfrac{8k}{9} = \dfrac{1012k}{315} = 1$ gives $k = \dfrac{315}{1012}$

b $P(R \leqslant 4) = P(R = 1 \text{ or } 3) = \dfrac{105}{506} + \dfrac{126}{506} = \dfrac{21}{46}$

EXERCISE 6B

1 $E(X) = \Sigma xp = 0 \times 0.1 + 1 \times 0.12 + 2 \times 0.36 + 3 \times 0.42 = 2.1$

$\text{Var}(X) = \Sigma x^2 p - \{E(X)\}^2 = 0^2 \times 0.1 + 1^2 \times 0.12 + 2^2 \times 0.36 + 3^2 \times 0.42 - 2.1^2 = 0.93$

Remember to subtract the square of the expectation when calculating the variance.

2 a Sum of probabilities is $0.3 + 2p + 0.32 + p + 0.05 = 1$, which gives $p = 0.2$.

  b $E(Y) = 0 \times 0.03 + 1 \times 0.4 + 2 \times 0.32 + 3 \times 0.2 + 4 \times 0.05 = 1.84$
  $SD(Y) = \sqrt{0^2 \times 0.03 + 1^2 \times 0.4 + 2^2 \times 0.32 + 3^2 \times 0.2 + 4^2 \times 0.05 - 1.84^2} = 0.946$

3 $P(T = 1) = P(T = 3) = P(T = 6) = P(T = 10) = \frac{1}{4}$
  $E(T) = \frac{1}{4} \times (1 + 3 + 6 + 10) = 5$; $Var(T) = \frac{1}{4} \times (1^2 + 3^2 + 6^2 + 10^2) - 5^2 = 11.5$

4 $E(V) = 1 \times 0.4 + 3 \times 0.28 + 9 \times 0.14 + 0.18m = 5.38$
  $2.5 + 0.18m = 5.38$, so $m = 16$
  $Var(V) = 1^2 \times 0.4 + 3^2 \times 0.28 + 9^2 \times 0.14 + 16^2 \times 0.18 - 5.38^2 = 31.3956$ or 31.4

5 $E(R) = 10 \times 0.05 + 20 \times 0.1 + 70 \times 0.35 + 100 \times 0.5 = 77$
  $Var(R) = 10^2 \times 0.05 + 20^2 \times 0.1 + 70^2 \times 0.35 + 100^2 \times 0.5 - 77^2 = 831$

6 $E(W) = 0.6 + 2.1 + 0.1a + 7.2 = a$, which gives $a = 11$
  $Var(W) = 2^2 \times 0.3 + 7^2 \times 0.3 + 11^2 \times 0.1 + 24^2 \times 0.3 - 11^2 = 79.8$

7 a $E(\text{grade}) = 5 \times 0.24 + 4 \times 0.33 + 3 \times 0.24 + 2 \times 0.11 + 1 \times 0.08 = 3.54$
  Expected outcome is 'a smallish profit' (half-way between 'no loss' and 'fair profit').
  $SD(\text{grade}) = \sqrt{5^2 \times 0.24 + 4^2 \times 0.33 + 3^2 \times 0.24 + 2^2 \times 0.11 + 1^2 \times 0.08 - 3.54^2} = 1.20$
  $SD(\text{grade})$ measures the variability of the profit.

  b $E(\text{grade}) = 2.46$; $SD(\text{grade}) = 1.20$; SD and expected outcome are both unchanged.

  > Although $E(\text{grade})$ changes from 3.54 to 2.46, this describes the same outcome as in part **a** (note that $5 - 3.54 = 2.46$).

8 a The grid shows lowest common multiples (LCMs).

6	6	6	6	12	30	6
5	5	10	15	20	5	30
4	4	4	12	4	20	12
3	3	6	3	12	15	6
2	2	2	6	4	10	6
1	1	2	3	4	5	6
	1	2	3	4	5	6

x	1	2	3	4	5	6	10	12	15	20	30
$P(X = x)$	$\frac{1}{36}$	$\frac{3}{36}$	$\frac{3}{36}$	$\frac{5}{36}$	$\frac{3}{36}$	$\frac{9}{36}$	$\frac{2}{36}$	$\frac{4}{36}$	$\frac{2}{36}$	$\frac{2}{36}$	$\frac{2}{36}$

b $E(X) = \frac{1}{36} \times (1 + 6 + 9 + 20 + 15 + 54 + 20 + 48 + 30 + 40 + 60) = 8\frac{5}{12}$

$P[X > E(X)] = P\left(X > 8\frac{5}{12}\right) = P(X = 10, 12, 15, 20 \text{ or } 30) = \frac{12}{36} = \frac{1}{3}$

c $\text{Var}(X) = \frac{4345}{36} - \left(8\frac{5}{12}\right)^2 = 49\frac{41}{48}$ or 49.9

9 a $P(H = 0) = 0.7 \times 0.7 \times 0.7 = 0.343$

$P(H = 1) = (0.3 \times 0.7 \times 0.7) + (0.7 \times 0.3 \times 0.7) + (0.7 \times 0.7 \times 0.3) = 0.441$

$P(H = 2) = (0.3 \times 0.3 \times 0.7) + (0.3 \times 0.7 \times 0.3) + (0.7 \times 0.3 \times 0.3) = 0.189$

$P(H = 3) = 0.3 \times 0.3 \times 0.3 = 0.027$

h	0	1	2	3
P(H = h)	0.343	0.441	0.189	0.027

b $E(H) = 0 \times 0.343 + 1 \times 0.441 + 2 \times 0.189 + 3 \times 0.027 = 0.9$

Expected number of hits in 1000 games is $1000 \times 0.9 = 900$.

10 a Let G be the number of girls selected.

$P(G = 0) = \frac{^{12}C_0 \times {^{18}C_2}}{^{30}C_2} = \frac{51}{145}$ $P(G = 1) = \frac{^{12}C_1 \times {^{18}C_1}}{^{30}C_2} = \frac{72}{145}$ $P(G = 2) = \frac{^{12}C_2 \times {^{18}C_0}}{^{30}C_2} = \frac{22}{145}$

$E(G) = 0 \times \frac{51}{145} + 1 \times \frac{72}{145} + 2 \times \frac{22}{145} = 0.8$ and $E(B) + E(G) = 2$, so $E(B) = 1.2$

b Ratio is 2 : 3; this is the same as the ratio for the number of girls to boys in the class.

c $\text{Var}(G) = 0^2 \times \frac{51}{145} + 1^2 \times \frac{72}{145} + 2^2 \times \frac{22}{145} - 0.8^2 = 0.463$ or $\frac{336}{725}$

11 a Let the number of yellow reels selected be Y.

$P(Y = 0) = \frac{^1C_0 \times {^7C_3}}{^8C_3} = 0.625$ and $P(Y = 1) = \frac{^1C_1 \times {^7C_2}}{^8C_3} = 0.375$

$E(Y) = 0 \times 0.625 + 1 \times 0.375 = 0.375$

b Let the number of red reels selected be R.

$P(R = 0) = \frac{^3C_0 \times {^5C_3}}{^8C_3} = \frac{10}{56}$ $P(R = 1) = \frac{^3C_1 \times {^5C_2}}{^8C_3} = \frac{30}{56}$

$P(R = 2) = \frac{^3C_2 \times {^5C_1}}{^8C_3} = \frac{15}{56}$ $P(R = 3) = \frac{^3C_3 \times {^5C_0}}{^8C_3} = \frac{1}{56}$

$E(R) = 0 \times \frac{10}{56} + 1 \times \frac{30}{56} + 2 \times \frac{15}{56} + 3 \times \frac{1}{56} = 1.125$ or $1\frac{1}{8}$

c $E(G) + E(Y) + E(R) = 3$, so $E(G) = 3 - 0.375 - 1.125 = 1.5$ or $1\frac{1}{2}$

12 a Expected profit is made from a combination of fixed fees (x) and claims of $540.

$0.7x + (0.3 \times 540) = 400$ gives $x =$ fixed fee $= \$340$.

b If the successful repayment rate is r, profit is $340r + 540(1 - r) > 400$
$$-200r > -140$$
$$r < 0.7 = 70\%$$

E(profit) > 40% if the successful repayment rate is below 70%.

13 a $P(X=0) = \frac{12}{20} \times \frac{12}{20} \times \frac{12}{20} = \frac{27}{125}$ $\quad P(X=1) = 3 \times \frac{8}{20} \times \frac{12}{20} \times \frac{12}{20} = \frac{54}{125}$

$P(X=2) = 3 \times \frac{8}{20} \times \frac{8}{20} \times \frac{12}{20} = \frac{36}{125}$ $\quad P(X=3) = \frac{8}{20} \times \frac{8}{20} \times \frac{8}{20} = \frac{8}{125}$

x	0	1	2	3
$P(X=x)$	$\frac{27}{125}$	$\frac{54}{125}$	$\frac{36}{125}$	$\frac{8}{125}$

$E(X) = \frac{1}{125} \times (0 \times 27 + 1 \times 54 + 2 \times 36 + 3 \times 8) = 1.2$

b The ratios $3 : 1.2$ and $n : 14$ are equal so $\frac{n}{14} = \frac{3}{1.2}$, which gives $n = 35$.

> Alternatively, $E(X) = n \times P(\text{junior selected})$,
> so $14 = n \times \frac{8}{20}$ which gives $n = 35$.

14 a $S \in \{1, 2, 3, 5\}$.

$P(S=1) = \frac{1}{6} + \left(\frac{3}{6} \times \frac{2}{6}\right) = \frac{1}{3}$ $\quad P(S=2) = \frac{3}{6} \times \frac{1}{6} = \frac{1}{12}$

$P(S=3) = \frac{1}{6} + \left(\frac{3}{6} \times \frac{2}{6}\right) = \frac{1}{3}$ $\quad P(S=5) = \frac{1}{6} + \left(\frac{3}{6} \times \frac{1}{6}\right) = \frac{1}{4}$

s	1	2	3	5
$P(S=s)$	$\frac{4}{12}$	$\frac{1}{12}$	$\frac{4}{12}$	$\frac{3}{12}$

b $E(S) = \frac{1}{12} \times (4 + 2 + 12 + 15) = 2\frac{3}{4}$ or 2.75

$P[S > E(S)] = P(S = 3 \text{ or } 5) = \frac{4}{12} + \frac{3}{12} = \frac{7}{12}$

c $\text{Var}(S) = \frac{1}{12} \times (1^2 \times 4 + 2^2 \times 1 + 3^2 \times 4 + 5^2 \times 3) - \left(2\frac{3}{4}\right)^2 = 2\frac{17}{48}$ or $\frac{113}{48}$

> The exact value of Var(S) cannot be given using decimals.

15 a $\dfrac{4!}{2! \times 2!} = 6$ arrangements (or list *AABB*, *ABAB*, *ABBA*, *BAAB*, *BABA* and *BBAA*).

Equally likely because each has a probability equal to the product of $\dfrac{1}{4}, \dfrac{1}{4}, \dfrac{3}{4}$ and $\dfrac{3}{4}$.

b $P(X=0) = \dfrac{1}{4} \times \dfrac{1}{4} \times \dfrac{1}{4} \times \dfrac{1}{4} = \dfrac{1}{256}$ $\qquad P(X=1) = 4 \times \left(\dfrac{3}{4} \times \dfrac{1}{4} \times \dfrac{1}{4} \times \dfrac{1}{4}\right) = \dfrac{12}{256}$

$P(X=2) = 6 \times \left(\dfrac{3}{4} \times \dfrac{3}{4} \times \dfrac{1}{4} \times \dfrac{1}{4}\right) = \dfrac{54}{256}$ $\qquad P(X=3) = 4 \times \left(\dfrac{3}{4} \times \dfrac{3}{4} \times \dfrac{3}{4} \times \dfrac{1}{4}\right) = \dfrac{108}{256}$

$P(X=4) = \dfrac{3}{4} \times \dfrac{3}{4} \times \dfrac{3}{4} \times \dfrac{3}{4} = \dfrac{81}{256}$

x	0	1	2	3	4
$P(X=x)$	$\dfrac{1}{256}$	$\dfrac{12}{256}$	$\dfrac{54}{256}$	$\dfrac{108}{256}$	$\dfrac{81}{256}$

$E(X) = \dfrac{1}{256} \times (0 + 12 + 108 + 324 + 324) = 3$

$Var(X) = \dfrac{1}{256} \times (0^2 \times 1 + 1^2 \times 12 + 2^2 \times 54 + 3^2 \times 108 + 4^2 \times 81) - 3^2 = \dfrac{3}{4}$

$\dfrac{Var(X)}{E(X)} = \dfrac{3}{4} \div 3 = \dfrac{1}{4}$

c $\dfrac{1}{4}$ represents the probability of not obtaining *B* with each spin.

Chapter 7
The binomial and geometric distributions

EXERCISE 7A

1 a $P(X=4) = \binom{4}{4} \times 0.2^4 \times 0.8^0 = 0.0016$

  b $P(X=0) = \binom{4}{0} \times 0.2^0 \times 0.8^4 = 0.4096$

  c $P(X=3) = \binom{4}{3} \times 0.2^3 \times 0.8^1 = 0.0256$

  d $P(X=3 \text{ or } 4) = 0.0256 + 0.0016 = 0.0272$

2 a $P(Y=7) = \binom{7}{7} \times 0.6^7 \times 0.4^0 = 0.6^7 = 0.0280$

  b $P(Y=5) = \binom{7}{5} \times 0.6^5 \times 0.4^2 = 0.261$

  c $P(Y \neq 4) = 1 - P(Y=4) = \binom{7}{4} \times 0.6^4 \times 0.4^3 = 0.710$

  d $P(3 < Y < 6) = P(Y=4 \text{ or } 5)$
  $= \binom{7}{4} \times 0.6^4 \times 0.4^3 + \binom{7}{5} \times 0.6^5 \times 0.4^2$
  $= 0.29030... + 0.26127... = 0.552$

> Premature rounding here will lead to an incorrect final answer of $0.290 + 0.261 = 0.551$.

3 a $P(W=5) = \binom{9}{5} \times 0.32^5 \times 0.68^4 = 0.0904$

  b $P(W \neq 5) = 1 - P(W=5) = 1 - 0.090397... = 0.910$

  c $P(W<2) = P(W=0 \text{ or } 1) = \binom{9}{0} \times 0.32^0 \times 0.68^9 + \binom{9}{1} \times 0.32^1 \times 0.68^8$
  $= 0.03108... + 0.13166... = 0.163$

  d $P(0 < W < 9) = 1 - P(W=0 \text{ or } 9) = 1 - [0.68^9 + 0.32^9] = 1 - 0.031122... = 0.969$

> In n trials, $P(\text{no successes}) = q^n$ and $P(n \text{ successes}) = p^n$.

4 a $P(V=4) = \binom{8}{4}\left(\frac{2}{7}\right)^4\left(\frac{5}{7}\right)^4 = 0.121$

b $P(V \geq 7) = P(V=7 \text{ or } 8) = \binom{8}{7}\left(\frac{2}{7}\right)^7\left(\frac{5}{7}\right)^1 + \left(\frac{2}{7}\right)^8 = 0.000933$

c $P(V \leq 2) = P(V=0, 1 \text{ or } 2) = \left(\frac{5}{7}\right)^8 + \binom{8}{1}\left(\frac{2}{7}\right)^1\left(\frac{5}{7}\right)^7 + \binom{8}{2}\left(\frac{2}{7}\right)^2\left(\frac{5}{7}\right)^6 = 0.588$

d $P(3 \leq V < 6) = P(V=3, 4 \text{ or } 5) = \binom{8}{3}\left(\frac{2}{7}\right)^3\left(\frac{5}{7}\right)^5 + \binom{8}{4}\left(\frac{2}{7}\right)^4\left(\frac{5}{7}\right)^4 + \binom{8}{5}\left(\frac{2}{7}\right)^5\left(\frac{5}{7}\right)^3 = 0.403$

e $P(V \text{ is odd}) = P(V=1, 3, 5 \text{ or } 7)$

$= \binom{8}{1}\left(\frac{2}{7}\right)^1\left(\frac{5}{7}\right)^7 + \binom{8}{3}\left(\frac{2}{7}\right)^3\left(\frac{5}{7}\right)^5 + \binom{8}{5}\left(\frac{2}{7}\right)^5\left(\frac{5}{7}\right)^3 + \binom{8}{7}\left(\frac{2}{7}\right)^7\left(\frac{5}{7}\right)^1 = 0.499$

5 a $P(\text{five heads}) = \binom{9}{5}\left(\frac{1}{2}\right)^5\left(\frac{1}{2}\right)^4 = \binom{9}{5}\left(\frac{1}{2}\right)^9 = 0.246$

b $P(\text{two sixes}) = \binom{11}{2}\left(\frac{1}{6}\right)^2\left(\frac{5}{6}\right)^9 = 0.296$

6 $P(\text{one brown}) = \binom{5}{1}\left(\frac{3}{4}\right)^1\left(\frac{1}{4}\right)^4 = 0.0146$

7 $P(\text{five pass}) = \binom{8}{5} \times 0.7^5 \times 0.3^3 = 0.254$

8 a Using $p = 0.63$ and $q = 0.37$, $P(20 \text{ male owners}) = \binom{30}{20} \times 0.63^{20} \times 0.37^{10} = 0.140$

b Using $p = 0.37$ and $q = 0.63$, $P(20 \text{ female owners}) = \binom{30}{20} \times 0.37^{20} \times 0.63^{10} = 0.000684$

9 $P(12 \text{ married}) = \binom{20}{12} \times 0.58^{12} \times 0.42^8 = 0.177$

10 a $P(\text{scores next ten}) = 0.95^{10} = 0.599$

b $P(\text{fails to score from one}) = \binom{7}{1} \times 0.05^1 \times 0.95^6 = 0.257$

11 $P(34 \text{ or } 35 \text{ succeed}) = \binom{40}{34} \times 0.87^{34} \times 0.13^6 + \binom{40}{35} \times 0.87^{35} \times 0.13^5 = 0.349$

12 a $P(\text{two days}) = \binom{14}{2} \times 0.15^2 \times 0.85^{12} = 0.291$

b $P(\text{at most two days}) = 0.85^{14} + \binom{14}{1} \times 0.15^1 \times 0.85^{13} + \binom{14}{2} \times 0.15^2 \times 0.85^{12} = 0.648$

13 a $P(\text{exactly one}) = \binom{200}{1} \times 0.003^1 \times 0.997^{199} = 0.330$

b $P(\text{fewer than two}) = 0.997^{200} + \binom{200}{1} \times 0.003^1 \times 0.997^{199} = 0.878$

Chapter 7: The binomial and geometric distributions

14 a P(one dropped) $= \binom{5}{1} \times 0.5^1 \times 0.5^4 = 5 \times 0.5^5 = 0.15625$ or $\frac{5}{32}$

b P(exactly one in at most one group) $= 0.84375^9 + \binom{9}{1} \times 0.15625^1 \times 0.84375^8 = 0.578$

15 $\dfrac{P(H=7)}{P(T=7)} = \left[\binom{12}{7}\left(\dfrac{3}{4}\right)^7\left(\dfrac{1}{4}\right)^5\right] \div \left[\binom{12}{7}\left(\dfrac{1}{4}\right)^7\left(\dfrac{3}{4}\right)^5\right] = \left(\dfrac{3}{4}\right)^{7-5} \times \left(\dfrac{1}{4}\right)^{5-7} = \left(\dfrac{3}{4}\right)^2 \times \left(\dfrac{1}{4}\right)^{-2} = 9$

16 $P(Q=0) = \binom{n}{0} \times 0.3^0 \times 0.7^n = 0.7^n$

Now $0.7^n > 0.1$

$n \log 0.7 > \log 0.1$

$n < \dfrac{\log 0.1}{\log 0.7} = 6.455\ldots$, so the greatest possible n is 6.

> The inequality sign must be reversed when multiplying or dividing throughout by a negative number, such as log 0.7.

17 $P(T=n) = \binom{n}{n} \times 0.96^n \times 0.04^0 = 0.96^n$

Now $0.96^n > 0.5$

$n \log 0.96 > \log 0.5$

$n < \dfrac{\log 0.5}{\log 0.96} = 16.979\ldots,$

so the greatest n is 16.

18 $P(R > n-1) = P(R=n) = \binom{n}{n} \times 0.8^n \times 0.2^0 = 0.8^n$

Now $0.8^n < 0.006$

$n \log 0.8 < \log 0.006$

$n > \dfrac{\log 0.006}{\log 0.8} = 22.926\ldots,$

so the least n is 23.

19 a $\binom{6}{0} \times p^0 \times (1-p)^6 = \dfrac{141\,393}{150\,000}$ gives $p = 1 - \sqrt[6]{0.94262} = 0.0098$

b $a = 150\,000 \times \binom{6}{2} \times 0.0098^2 \times 0.9902^4 = 208$; $b = 150\,000 - (141\,393 + 8396 + 208) = 3$.

c $(150\,000 + n) \times \binom{6}{1} \times 0.0098^1 \times 0.9902^5 \geq 8400$

$n \geq \dfrac{8400}{0.05597\ldots} - 150\,000 = 67.742\ldots$, so the least number of additional cartons is 68.

20 a P(no months with more than 5 m) = $(1-p)^4 = \frac{2}{32}$ gives $p = \frac{1}{2}$ or 0.5.

p represents the probability of more than 5 metres of rainfall in any given month of the monsoon season.

b The probability of more than 5 metres of rainfall in any given month of the monsoon season is unlikely to be constant *or*

Whether one month has more than 5 metres of rainfall is unlikely to be independent of whether another month has.

21 a $0.9^4 = 0.6561$

b

	Drinks tea	
	with sugar	without sugar
Male	$0.9 \times 0.6 = 0.54$	$0.9 \times 0.4 = 0.36$
Female	$0.9 \times 0.4 = 0.36$	$0.9 \times 0.6 = 0.54$

P(2M and 2F) + P(1M and 1F) + P(0M and 0F)
= $(0.54^2 \times 0.36^2) + (2 \times 0.36 \times 0.54)(2 \times 0.54 \times 0.36) + (0.36^2 \times 0.54^2) = 0.227$

22 P(0LH and 1RH) = $0.995^{200} \times \binom{300}{1} \times 0.004^1 \times 0.996^{299} = 0.13284...$

P(1LH and 0RH) = $\binom{200}{1} \times 0.005^1 \times 0.995^{199} \times 0.996^{300} = 0.11081...$

P(exactly 1 colour-blind) = $0.13284... + 0.11081... = 0.244$

EXERCISE 7B

1 a V has $n = 5$, $p = 0.2$ and $q = 0.8$.
$E(V) = np = 5 \times 0.2 = 1$; $Var(V) = npq = 5 \times 0.2 \times 0.8 = 0.8$; $SD(V) = \sqrt{0.8} = 0.894$

b W has $n = 24$, $p = 0.55$ and $q = 0.45$.
$E(W) = np = 24 \times 0.55 = 13.2$; $Var(W) = npq = 24 \times 0.55 \times 0.45 = 5.94$; $SD(W) = \sqrt{5.94} = 2.44$

c X has $n = 365$, $p = 0.18$ and $q = 0.82$.
$E(X) = np = 365 \times 0.18 = 65.7$; $Var(X) = npq = 365 \times 0.18 \times 0.82 = 53.874$; $SD(X) = \sqrt{57.874} = 7.34$

d Y has $n = 20$, $p = \sqrt{0.5}$ and $q = 1 - \sqrt{0.5}$.
$E(Y) = np = 20 \times \sqrt{0.5} = 14.1$; $Var(Y) = npq = 20 \times \sqrt{0.5} \times (1 - \sqrt{0.5}) = 4.14$; $SD(Y) = \sqrt{4.142...} = 2.04$

2 a $E(X) = 8 \times 0.25 = 2$; $Var(X) = 8 \times 0.25 \times 0.75 = 1.5$

b $P[X = E(X)] = P(X = 2) = \binom{8}{2} \times 0.25^2 \times 0.75^6 = 0.311$

c $P[X < E(X)] = P(X < 2) = P(X = 0) + P(X = 1) = 0.75^8 + \binom{8}{1} \times 0.25^1 \times 0.75^7 = 0.367$

Chapter 7: The binomial and geometric distributions

3 **a** $P(Y \neq 3) = 1 - P(Y = 3) = 1 - \binom{11}{3} \times 0.23^3 \times 0.77^8 = 0.752$

   b $E(Y) = 11 \times 0.23 = 2.53$.
   $P[Y < E(X)] = P(Y < 2.53) = P(Y = 0, 1 \text{ or } 2)$
   $= 0.77^{11} + \binom{11}{1} \times 0.23^1 \times 0.77^{10} + \binom{11}{2} \times 0.23^2 \times 0.77^9 = 0.519$

4 **a** $\dfrac{\text{Var}(X)}{E(X)} = q = \dfrac{12}{20}$, so $p = \dfrac{8}{20}$, and $n \times \dfrac{8}{20} = 20$ gives $n = 50$.

   $\dfrac{\text{Var}(X)}{E(X)} = \dfrac{npq}{np} = q$

   b $P(X = 21) = \binom{50}{21} \times 0.4^{21} \times 0.6^{29} = 0.109$

5 **a** $\dfrac{\text{Var}(G)}{E(G)} = q = 10\dfrac{5}{24} \div 24\dfrac{1}{2} = \dfrac{5}{12}$, so $p = \dfrac{7}{12}$, and $n \times \dfrac{7}{12} = 24\dfrac{1}{2}$ gives $n = 42$.

   b $P(G = 20) = \binom{42}{20} \times \left(\dfrac{7}{12}\right)^{20} \times \left(\dfrac{5}{12}\right)^{22} = 0.0462$

6 $\dfrac{\text{Var}(W)}{E(W)} = q = \dfrac{0.27}{2.7} = 0.1$, $p = 0.9$, and $n = \dfrac{np}{p} = \dfrac{E(W)}{p} = \dfrac{2.7}{0.9} = 3$.

   $W \in \{0, 1, 2, 3\}$ and $P(W = w) = \binom{3}{w} \times 0.9^w \times 0.1^{3-w}$.

w	0	1	2	3
P(W = w)	0.001	0.027	0.243	0.729

7 **a** E.g. X is not a discrete variable *or* There are more than two possible outcomes.

   b E.g. Selections are not independent.

   c E.g. X can take the value 0 only *or* X is not a variable.

8 $E(Q) = \dfrac{n}{3}$, $\text{Var}(Q) = \dfrac{2n}{9}$ and $SD(Q) = \dfrac{\sqrt{2n}}{3}$, so $\dfrac{\sqrt{2n}}{3} = \dfrac{1}{3} \times \dfrac{n}{3}$ which gives $n = 18$.

   $P(5 < Q < 8) = P(Q = 6 \text{ or } 7) = \binom{18}{6}\left(\dfrac{1}{3}\right)^6\left(\dfrac{2}{3}\right)^{12} + \binom{18}{7}\left(\dfrac{1}{3}\right)^7\left(\dfrac{2}{3}\right)^{11} = 0.364$

9 $E(H) = 192p$ and $SD(H) = \sqrt{192p(1-p)}$

   $192p = 24 \times \sqrt{192p(1-p)}$ becomes $256p^2 - 192p = 0$, giving $p = \dfrac{3}{4}$.

   $k \times 2^{-379} = \binom{192}{2}\left(\dfrac{3}{4}\right)^2\left(\dfrac{1}{4}\right)^{190}$ which gives $k = \dfrac{18336 \times 3^2 \times 2^{379}}{2^4 \times 2^{380}} = 5157$.

10 **a** $462 \times 0.013 = 6.006$

   b $\text{Var(damaged)} = 462 \times 0.013 \times (1 - 0.013) = 5.93$
   $\text{Var(undamaged)} = 462 \times (1 - 0.013) \times 0.013 = 5.93$

c $\binom{462}{8} \times 0.013^8 \times 0.987^{454} = 0.1039... \approx 10.4\%$

d P(at least one) = 1 − P(none) = $1 - 0.896083396^2 = 0.197$

11 a $50 \times 0.92 = 46$

b $50 \times 0.08 \times 0.92 = 3.68$

c i $\binom{50}{3} \times 0.08^3 \times 0.92^{47} + \binom{50}{4} \times 0.08^4 \times 0.92^{46} + \binom{50}{5} \times 0.08^5 \times 0.92^{45} = 0.566$

ii $\binom{2}{2} \times 0.565899438^2 = 0.320$

EXERCISE 7C

1 a X has $p = 0.2$, $q = 0.8$, so
$P(X = 7) = pq^6 = 0.2 \times 0.8^6 = 0.0524$

b $P(X \neq 5) = 1 - P(X = 5) = 1 - pq^4$
$= 1 - (0.2 \times 0.8^4) = 0.91808$

c $P(X > 4) = q^4 = 0.8^4 = 0.4096$

> Alternatively, use
> $P(X > 4) = 1 - P(X \leq 4)$
> $= 1 - P(X = 0, 1, 2, 3 \text{ or } 4)$.

2 a $P(T = 3) = pq^2 = 0.32 \times 0.68^2 = 0.148$

b $P(T \leq 6) = 1 - q^6 = 1 - 0.68^6 = 0.901$

c $P(T > 7) = q^7 = 0.68^7 = 0.0672$

3 a $P(X = 3) = pq^2 = 0.5 \times 0.5^2 = 0.125$

b $P(X < 4) = P(X \leq 3) = 1 - q^3 = 1 - 0.5^3 = 0.875$

4 a $P(X = 8) = \left(\frac{1}{6}\right)\left(\frac{5}{6}\right)^7 = 0.0465$

b $P(X > 4) = \left(\frac{5}{6}\right)^4 = 0.482$

5 a $P(X = 2) = 0.4 \times 0.6 = 0.24$

b $P(X \leq 5) = 1 - 0.6^5 = 0.922$

c $P(X \geq 8) = P(X > 7) = 0.6^7 = 0.0280$

6 a i $P(X = 3) = 0.8 \times 0.2^2 = 0.032$

ii $P(X > 4) = 0.2^4 = 0.0016$

b $p = 0.8 \times 0.9 = 0.72$, $q = 0.28$, and
P(second customer) $= 0.72 \times 0.28 = 0.2016$.

7 a i $P(X = 12) = 0.07 \times 0.93^{11} = 0.0315$

ii $P(X > 10) = 0.93^{10} = 0.484$

iii $P(X \leq 8) = 1 - 0.93^8 = 0.440$

b Faults occur independently and at random.

8 a $P(X = 2) = 0.3 \times 0.7 = 0.21$

b $P(Y = 2) = 0.7 \times 0.3 = 0.21$

c $P(X = 1 \text{ and } Y = 1) = P(X = 1) \times P(Y = 1)$
$= 0.3 \times 0.7 = 0.21$

9 a $P(X \leq 3) = 1 - 0.86^3 = 0.364$

b $P(X \geq 5) = P(X > 4) = 0.86^4 = 0.547$

10 $P(X = 5) = 0.44 \times 0.56^4 = 0.0433$

11 a Not suitable; trials are not identical (p is not constant).

b Not suitable; success is dependent on the previous two letters typed *or* X cannot be equal to 1 or 2 *or* p is not constant.

c It is suitable.

d Not suitable; trials are not identical *or* p is not constant *or* He may win none.

Chapter 7: The binomial and geometric distributions

12 $\dfrac{P(T=2)}{P(T=5)} = \dfrac{p(1-p)}{p(1-p)^4} = \dfrac{125}{8}$ gives $p = 1 - \sqrt[3]{\dfrac{8}{125}} = 0.6$, so $P(T=3) = 0.6 \times 0.4^2 = 0.096$

13 $0.2464 = p(1-p)$ becomes $p^2 - p + 0.2464 = 0$
$(p - 0.44)(p - 0.56) = 0$, so $p = 0.44$ (as we know $p < 0.5$).
$P(X > 3) = (1 - 0.44)^3 = 0.176$

14 $1 - q^4 = \dfrac{2385}{2401}$ gives $q = 1 - \sqrt[4]{\dfrac{2385}{2401}} = \dfrac{2}{7}$ and $p = \dfrac{5}{7}$

$P(1 \leqslant X < 4) = P(X = 1, 2 \text{ or } 3) = \dfrac{5}{7} + \left(\dfrac{5}{7} \times \dfrac{2}{7}\right) + \left(\dfrac{5}{7} \times \dfrac{2}{7} \times \dfrac{2}{7}\right) = \dfrac{335}{343}$ or 0.977

15 a $P(\text{double}) = 6 \times \left(\dfrac{1}{6}\right)^2 = \dfrac{1}{6}$, so $P(X = 4) = \left(\dfrac{1}{6}\right)\left(\dfrac{5}{6}\right)^3 = 0.0965$ or $\dfrac{125}{1296}$

b $P(\text{sum} > 10) = 3 \times \left(\dfrac{1}{6}\right)^2 = \dfrac{1}{12}$, so $P(X < 10) = P(X \leqslant 9) = 1 - \left(1 - \dfrac{1}{12}\right)^9 = 0.543$

16 There are three ways in which the sum of X and Y can be equal to 4.

	X	Y	Probability
	1	3	$0.24 \times (0.25 \times 0.75^2) = 0.03375$
or	2	2	$(0.24 \times 0.76) \times (0.25 \times 0.75) = 0.0342$
or	3	1	$(0.24 \times 0.76^2) \times 0.25 = 0.034656$
			Total $= 0.102606$

$P(X + Y = 4) = 0.103$

EXERCISE 7D

1 $p = 0.36 = \dfrac{9}{25}$, so $E(X) = \dfrac{1}{p} = \dfrac{25}{9}$ or $2\dfrac{7}{9}$

2 $P(Y = 1) = p$, so $p = 0.2$, and $E(Y) = \dfrac{1}{p} = \dfrac{1}{0.2} = 5$

3 $E(S) = \dfrac{1}{p} = \dfrac{9}{2}$ gives $p = \dfrac{2}{9}$ and $q = \dfrac{7}{9}$, so
$P(S = 2) = pq = \dfrac{2}{9} \times \dfrac{7}{9} = \dfrac{14}{81}$

4 Mode is 1; mean $= \dfrac{1}{p} = \dfrac{1}{0.5} = 2$

The mode of all geometric distributions is 1.

5 $E(X) = \dfrac{1}{p} = 1 \div \dfrac{1}{6} = 6$, and
$P(X > 6) = \left(1 - \dfrac{1}{6}\right)^6 = 0.335$

6 a $P(\text{non-prime}) = P(1) = p = \dfrac{1}{16}$, so
$E(X) = 1 \div \dfrac{1}{16} = 16$

b $P(\text{prime}) = \dfrac{15}{16}$, so
$P(X = 3) = \dfrac{15}{16} \times \left(\dfrac{1}{16}\right)^2 = \dfrac{15}{4096}$ or 0.00366

7 a $E(\text{Thierry}) = 1 \div \dfrac{5}{8} = 1.6$ and
$E(\text{Sylvie}) = 1 \div \dfrac{4}{7} = 1.75$
Thierry is expected to fail fewer times.

b $\left(\dfrac{4}{7} \times \dfrac{3}{7}\right) \times \left(\dfrac{5}{8} \times \dfrac{3}{8}\right) = \dfrac{45}{784}$ or 0.0574

8 **a** Cards are selected with replacement, so that selections are independent.

b i P(diamond) = $\frac{1}{4}$, so E(X) = 4, and

$$P(X=4) = \frac{1}{4} \times \left(\frac{3}{4}\right)^3 = \frac{27}{256} \text{ or } 0.105$$

ii Four ways: SSD, SCD, CSD and CCD.

Probability is $4 \times 4 \times \left(\frac{1}{4}\right)^3 = \frac{1}{16}$ or 0.0625

9 E(X) = $\frac{1}{0.002}$ = 500, and P(X ⩽ b) = 1 − 0.998^b.

Now $1 - 0.998^b > 0.865$

$0.998^b < 0.135$

$b \log 0.998 < \log 0.135$

$b > \frac{\log 0.135}{\log 0.998} = 1000.23...,$

so the smallest possible b is 1001.

10 **a**

	1st toss	2nd toss	3rd toss
Anouar	tail	tail	head
Zane	tail	tail	

Sequence is A_{tail}, Z_{tail}, A_{tail}, Z_{tail}, A_{head}, i.e. two tails each then a head for Anouar.

b $0.5^2 + 0.5^4 + 0.5^6 + 0.5^8 + ...$

c $\frac{P(\text{Anouar wins})}{P(\text{Zane wins})} = \frac{0.5 + 0.5^3 + 0.5^5 + 0.5^7 + ...}{0.5^2 + 0.5^4 + 0.5^6 + 0.5^8 + ...} = \frac{0.5(1 + 0.5^2 + 0.5^4 + 0.5^6 + ...)}{0.5^2(1 + 0.5^2 + 0.5^4 + 0.5^6 + ...)} = \frac{1}{0.5} = 2.$

Anouar is twice as likely to win a game as Zane, so P(Anouar wins) = $\frac{2}{3}$.

Chapter 8
The normal distribution

EXERCISE 8A

1 a False (A is centred to the left of B).
  b True (the values in B are more spread out than the values in A).
  c False (same reason as part **b**).
  d False (same reason as part **b**).
  e True (more than half the area under curve B is to the right of μ_A).
  f False (more than half the area under curve A is to the left of μ_B).

> The centre of a normal curve is its axis of symmetry, which is where we find the mean, median and mode.

2 a i $\sigma_P > \sigma_Q$ (values in P are more widely spread out than values in Q).
   ii Median for P < median for Q (P is centred to the left of Q).
   iii IQR for P > IQR for Q (values in P are more widely spread out than values in Q).
  b i Range of W is the same as range of P (all values in Q are contained within the range of values in P).
   ii The probability distribution for W is not a normal curve.
       High values of W are more likely than low values (W is negatively skewed).
   iii $\mu_P < \mu_W < \mu_Q$ with μ_W to the left of the peak on curve W.

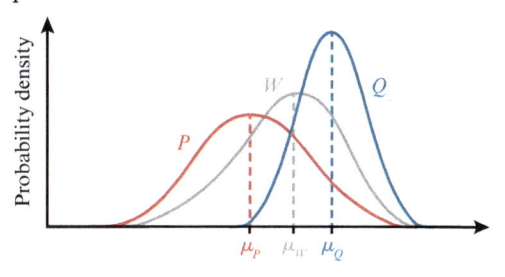

3 Areas under the three graphs are equal.
  For women and men together, the width of the graph spans the widths of the two original graphs and is centred on $\frac{160+180}{2} = 170$ cm.

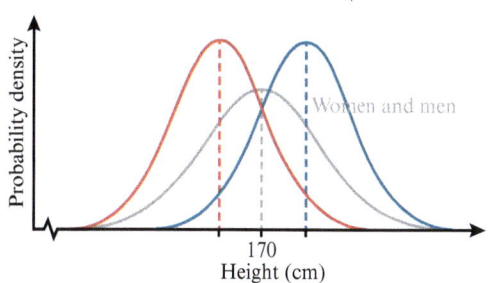

4 a

  b Peach juice curve is shorter and wider than apple juice curve.
  Both curves are symmetric, bell-shaped, centred on 340 ml, and they have equal areas.

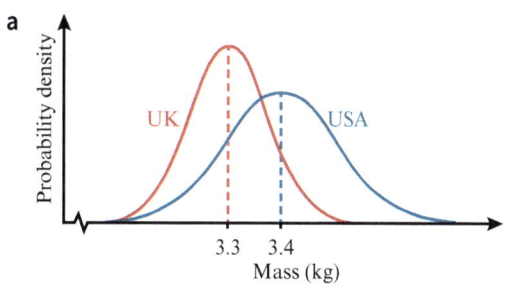

5 a

  b USA curve is shorter and wider and centred to the right of the UK curve.
  Both curves are symmetric, bell-shaped, and they have equal areas.

6 a $\mu_x = \dfrac{12000}{5000} = 2.4$ and $\mu_y = \dfrac{26000}{10000} = 2.6$, so $\mu_y > \mu_x$

b $\sigma_x = \sqrt{\dfrac{35000}{5000} - 2.4^2} = 1.11$ and $\sigma_y = \sqrt{\dfrac{72000}{10000} - 2.6^2} = 0.663$, so $\sigma_x > \sigma_y$

Curve for x is shorter and wider, and centred to the left of the curve for y.

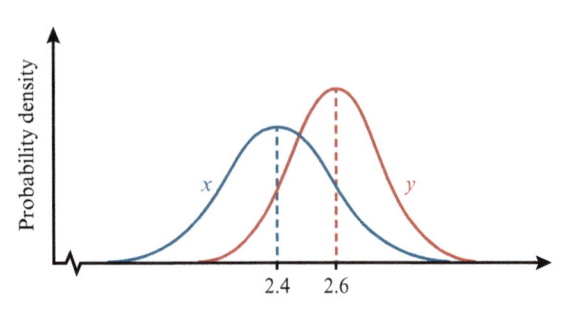

> We can compare the widths of normal curves for finite sets of data because they exist between lower and upper boundary values. Normally distributed random variables exist in the domain from $-\infty$ to $+\infty$ so, strictly speaking, they all have the same width.

EXERCISE 8B

1 a Area to the left of $z = 0.567$ is $\Phi(0.567) = 0.715$

b $\Phi(2.468) = 0.993$

c Area to the right of $z = -1.53$ is
$\Phi(1.53) = 0.937$

d $\Phi(0.077) = 0.531$

e Area to the right of $z = 0.817$ is
$1 - \Phi(0.817) = 0.207$

f $1 - \Phi(2.009) = 0.0224$

g Area to the left of $z = -1.75$ is
$1 - \Phi(1.75) = 0.0401$

h $1 - \Phi(0.013) = 0.495$

i $\Phi(1.96) = 0.975$

j $1 - \Phi(2.576) = 0.005$

> The area to the left of a positive z-value and to the right of a negative z-value is greater than 0.5.
> The area to the right of a positive z-value and to the left of a negative z-value is less than 0.5.

2 a $\Phi(2.5) - \Phi(1.5) = 0.9938 - 0.9332 = 0.0606$

b $\Phi(1.272) - \Phi(0.046) = 0.8984 - 0.5184 = 0.380$

c $\Phi(2.326) - \Phi(1.645) = 0.99 - 0.95 = 0.0400$

d $\Phi(2.807) - \Phi(1.282) = 0.9975 - 0.90 = 0.0975$

e $\Phi(1.777) - \Phi(0.746) = 0.9622 - 0.7722 = 0.190$

f $\Phi(1.008) - \Phi(0.337) = 0.8432 - 0.6319 = 0.211$

g $\Phi(1.2) - [1 - \Phi(1.2)] = 2\Phi(1.2) - 1$
$= 2 \times 0.8849 - 1 = 0.770$

h $\Phi(2.667) - [1 - \Phi(1.667)]$
$= 0.9962 - 1 + 0.9522 = 0.948$

i $\Phi(1.6) - [1 - \Phi(0.75)] = 0.9452 - 1 + 0.7734$
$= 0.719$

j $\Phi(2.236) - \Phi(1.414) = 0.9873 - 0.9213$
$= 0.066$

3 a $k = \Phi^{-1}(0.9087) = 1.333$ or 1.33

b $k = \Phi^{-1}(0.5442) = 0.111$

c $k = \Phi^{-1}(1 - 0.2743) = \Phi^{-1}(0.7257) = 0.600$

d $k = \Phi^{-1}(1 - 0.0298) = \Phi^{-1}(0.9702)$
$= 1.884$ or 1.88

e $-k = \Phi^{-1}(1 - 0.25) = \Phi^{-1}(0.75)$, so $k = -0.674$

f $-k = \Phi^{-1}(1 - 0.3552) = \Phi^{-1}(0.6448)$, so
$k = -0.371$

g $-k = \Phi^{-1}(0.9296)$, so $k = -1.473$ or -1.47

h $-k = \Phi^{-1}(0.648)$, so $k = -0.380$

i $\Phi(k) - 0.5 = \dfrac{0.9128}{2}$, so $\Phi(k) = 0.9564$
$k = \Phi^{-1}(0.9564) = 1.71$

j $\Phi(k) - 0.5 = 0.6994 \div 2$, so $\Phi(k) = 0.8497$
$k = \Phi^{-1}(0.8497) = 1.035$ or 1.04

4 a $\Phi(c) = \Phi(1.638) - 0.2673 = 0.6819$
$c = \Phi^{-1}(0.6819) = 0.473$

b $\Phi(c) = \Phi(2.878) - 0.4968 = 0.5012$
$c = \Phi^{-1}(0.5012) = 0.00300$

c $\Phi(c) = 0.1408 + \Phi(1) = 0.9821$
$c = \Phi^{-1}(0.9821) = 2.10$

d $\Phi(c) = 0.35 + \Phi(0.109) = 0.8934$
$c = \Phi^{-1}(0.8934) = 1.245$ or 1.25

e $\Phi(c) = \Phi(2) - 0.6687 = 0.3085$
$c = -\Phi^{-1}(1 - 0.3085) = -\Phi^{-1}(0.6915) = -0.500$

f $\Phi(c) = \Phi(1.85) - 0.9516$
$\Phi(c) = 0.0162$
$c = -\Phi^{-1}(1 - 0.0162) = -\Phi^{-1}(0.9838) = -2.139$ or -2.14

g $\Phi(c) - [1 - \Phi(1.221)] = 0.888$
$\Phi(c) = 0.999$, so $c = 3.09$

h $\Phi(c) - [1 - \Phi(0.674)] = 0.725$
$\Phi(c) = 0.975$, so $c = 1.96$

i $\Phi(c) - [1 - \Phi(2.63)] = 0.6861$
$\Phi(c) = 0.6904$, so $c = 0.497$

j $\Phi(2.7) - \Phi(-c) = 0.0252$
$\Phi(-c) = 0.9713$
$c = -\Phi^{-1}(0.9713) = -1.90$

EXERCISE 8C

1 a $P(X \leq 11) = \Phi\left(\dfrac{11 - 8}{\sqrt{25}}\right) = \Phi(0.6) = 0.726$

b $P(X < 69.1) = \Phi\left(\dfrac{69.1 - 72}{\sqrt{11}}\right) = 1 - \Phi(0.874) = 0.191$

c $P(3 < X < 7) = \Phi\left(\dfrac{7 - 5}{\sqrt{5}}\right) - \Phi\left(\dfrac{3 - 5}{\sqrt{5}}\right)$
$= \Phi(0.894) - [1 - \Phi(0.894)]$
$= 2\Phi(0.894) - 1$
$= 0.629$

> If a and b are equidistant on either side of the mean then $P(a < X < b) = 2[\Phi(z_b) - 0.5] = 2\Phi(z_b) - 1$, where z_b is the standardised value of b. This is because the area representing the probability is symmetric about the mean.

2 a $P(X \leq 9.7) = \Phi\left(\dfrac{9.7 - 6.2}{\sqrt{6.25}}\right) = \Phi(1.4) = 0.919$
$P(X > 9.7) = 1 - 0.9192 = 0.0808$

b $P(X \leq 5) = \Phi\left(\dfrac{5 - 3}{\sqrt{49}}\right) = \Phi(0.286) = 0.613$
$P(X > 5) = 1 - 0.613 = 0.387$

c $P(X > 33.4) = 1 - \Phi\left(\dfrac{33.4 - 37}{\sqrt{4}}\right) = 1 - \Phi(-1.8) = \Phi(1.8) = 0.964$
$P(X \leq 33.4) = 1 - 0.9641 = 0.0359$

d $P(X \geq 13.5) = 1 - \Phi\left(\dfrac{13.5 - 20}{\sqrt{15}}\right) = 1 - \Phi(-1.678) = \Phi(1.678) = 0.953$

$P(X < 13.5) = 1 - 0.9533 = 0.0467$

e $P(X \leq 91) = \Phi\left(\dfrac{91 - 80}{\sqrt{375}}\right) = \Phi(0.568) = 0.715$

$P(X > 91) = 1 - 0.715 = 0.285$

f $P(1 \leq X < 21) = \Phi\left(\dfrac{21 - 11}{\sqrt{25}}\right) - \Phi\left(\dfrac{1 - 11}{\sqrt{25}}\right) = \Phi(2) - \Phi(-2) = 2\Phi(2) - 1 = 0.954$

g $P(2 \leq X < 5) = \Phi\left(\dfrac{5 - 3}{\sqrt{7}}\right) - \Phi\left(\dfrac{2 - 3}{\sqrt{7}}\right) = \Phi(0.756) + \Phi(0.378) - 1 = 0.422$

h $P(6.2 \geq X \geq 8.8) = \Phi\left(\dfrac{6.2 - 7}{\sqrt{1.44}}\right) + 1 - \Phi\left(\dfrac{8.8 - 7}{\sqrt{1.44}}\right) = \Phi(-0.667) + 1 - \Phi(1.5)$

$= 1 - \Phi(0.667) + 1 - \Phi(1.5) = 2 - \Phi(1.5) - \Phi(0.667) = 0.319$

i $P(26 \leq X < 28) = \Phi\left(\dfrac{28 - 25}{\sqrt{6}}\right) - \Phi\left(\dfrac{26 - 25}{\sqrt{6}}\right) = \Phi(1.225) - \Phi(0.408) = 0.231$

j $P(8 \leq X < 10) = \Phi\left(\dfrac{10 - 12}{\sqrt{2.56}}\right) - \Phi\left(\dfrac{8 - 12}{\sqrt{2.56}}\right) = \Phi(-1.25) - \Phi(-2.5)$

$= 1 - \Phi(1.25) - [1 - \Phi(2.5)] = \Phi(2.5) - \Phi(1.25) = 0.0994$

3 a $\dfrac{a - 30}{\sqrt{16}} = \Phi^{-1}(0.8944)$ gives $a = 30 + 4 \times 1.25 = 35.0$

b $\dfrac{b - 12}{\sqrt{4}} = \Phi^{-1}(0.9599)$ gives $b = 12 + 2 \times 1.75 = 15.5$

c $\dfrac{23 - c}{\sqrt{9}} = \Phi^{-1}(0.9332)$ gives $c = 23 - 3 \times 1.5 = 18.5$

d $\dfrac{d - 17}{\sqrt{25}} = \Phi^{-1}(1 - 0.0951)$ gives $d = 17 + 5 \times 1.31 = 23.6$

e $\dfrac{100 - e}{\sqrt{64}} = \Phi^{-1}(0.95)$ gives $e = 100 - 8 \times 1.645 = 86.84$ or 86.8

4 a $P(f \leq X < 13.3) < P(\mu \leq X < 13.3)$, so $f > \mu$

$$\Phi\left(\dfrac{13.3 - 10}{\sqrt{7}}\right) - \Phi\left(\dfrac{f - 10}{\sqrt{7}}\right) = 0.1922$$

$$\Phi\left(\dfrac{f - 10}{\sqrt{7}}\right) = \Phi(1.247) - 0.1922 = 0.7016$$

$$\dfrac{f - 10}{\sqrt{7}} = \Phi^{-1}(0.7016)$$

$$f = 10 + 0.529 \times \sqrt{7} = 11.4$$

b $P(g \leqslant X < 55) > P(\mu \leqslant X < 55)$, so $g < \mu$

$$\Phi\left(\frac{55-45}{\sqrt{50}}\right) - \Phi\left(\frac{g-45}{\sqrt{50}}\right) = 0.5486$$

$$\Phi\left(\frac{g-45}{\sqrt{50}}\right) = \Phi(1.414) - 0.5486 = 0.3727$$

$$\frac{45-g}{\sqrt{50}} = \Phi^{-1}(1-0.3727) = \Phi^{-1}(0.6273) = 0.325$$

$$g = 45 - 0.325 \times \sqrt{50} = 42.7$$

c $\Phi\left(\frac{h-7}{\sqrt{2}}\right) - \Phi\left(\frac{8-7}{\sqrt{2}}\right) = 0.216$

$$\frac{h-7}{\sqrt{2}} = \Phi^{-1}(0.9761)$$

$$h = 7 + 1.98 \times \sqrt{2} = 9.80$$

d $\Phi\left(\frac{22-20}{\sqrt{11}}\right) - \Phi\left(\frac{j-20}{\sqrt{11}}\right) = 0.5$

$$\Phi\left(\frac{j-20}{\sqrt{11}}\right) = \Phi(0.603) - 0.5 = 0.2267$$

$$\frac{20-j}{\sqrt{11}} = \Phi^{-1}(1-0.2267) = \Phi^{-1}(0.7733) = 0.749$$

$$j = 20 - 0.749 \times \sqrt{11} = 17.5$$

5 $P(X < 0) = \Phi\left(\frac{0-4}{\sqrt{6}}\right) = 1 - \Phi\left(\frac{4-0}{\sqrt{6}}\right) = 1 - \Phi(1.633) = 0.0513$

6 $P(X < 2\mu) = \Phi\left(\dfrac{2\mu - \mu}{\frac{2}{3}\mu}\right) = \Phi(1.5) = 0.933$

7 $\Phi\left(\frac{14.7-10}{\sigma}\right) = 1 - 0.04 = 0.96$, so $\frac{4.7}{\sigma} = \Phi^{-1}(0.96) = 1.751$, giving $\sigma = 2.68$

8 $\Phi\left(\frac{15-\mu}{\sqrt{13}}\right) = 0.75$, so $\frac{15-\mu}{\sqrt{13}} = \Phi^{-1}(0.75) = 0.674$

$\mu = 15 - 0.674 \times \sqrt{13} = 12.6$

9 $\Phi\left(\frac{83-\mu}{\sigma}\right) = 0.95$, so $\frac{83-4\sigma}{\sigma} = \Phi^{-1}(0.95) = 1.645$

$5.645\sigma = 83$, so $\sigma = 14.7$ and $\mu = 58.8$

10 $\Phi\left(\frac{\mu-12}{\mu-30}\right) = 0.9$, so $\frac{\mu-12}{\mu-30} = \Phi^{-1}(0.9) = 1.282$

$0.282\mu = 30 \times 1.282 - 12$, so $\mu = 93.8$ and $\sigma = 63.8$

11 $\Phi\left(\dfrac{\mu - 1.288}{\sigma}\right) = 0.719$, so $\dfrac{\mu - 1.288}{\sigma} = \Phi^{-1}(0.719) = 0.58$, giving the equation

$\mu - 1.288 = 0.58\sigma$ .... [1]

$\Phi\left(\dfrac{6.472 - \mu}{\sigma}\right) = 0.591$, so $\dfrac{6.472 - \mu}{\sigma} = \Phi^{-1}(0.591) = 0.23$, giving the equation

$6.472 - \mu = 0.23\sigma$ .... [2]

Solving [1] and [2] simultaneously gives $\sigma = 6.4$ and $\mu = 5$

$P(4 \leqslant Q < 5) = 0.5 - \left[1 - \Phi\left(\dfrac{5 - 4}{6.40}\right)\right] = \Phi(0.156) - 0.5 = 0.0620$

12 $\Phi\left(\dfrac{8.4 - \mu}{\sigma}\right) = 0.7509$, so $\dfrac{8.4 - \mu}{\sigma} = \Phi^{-1}(0.7509) = 0.667$, giving the equation

$8.4 - \mu = 0.677\sigma$ .... [1]

$1 - \Phi\left(\dfrac{9.2 - \mu}{\sigma}\right) = 0.1385$, so $\dfrac{9.2 - \mu}{\sigma} = \Phi^{-1}(0.8615) = 1.087$, giving the equation

$9.2 - \mu = 1.087\sigma$ .... [2]

Solving [1] and [2] gives $\mu = 7.08$ and $\sigma = 1.95$ (to 3 significant figures).

$P(V \leqslant 10) = \Phi\left(\dfrac{10 - 7.0790\ldots}{1.9512\ldots}\right) = \Phi(1.497) = 0.933$

13 $\Phi\left(\dfrac{\mu - 4.75}{\sigma}\right) = 0.6858$, so $\dfrac{\mu - 4.75}{\sigma} = \Phi^{-1}(0.6858) = 0.484$, giving the equation

$\mu - 4.75 = 0.484\sigma$ .... [1]

$1 - \Phi\left(\dfrac{\mu - 2.25}{\sigma}\right) = 0.0489$, so $\dfrac{\mu - 2.25}{\sigma} = \Phi^{-1}(0.9511) = 1.656$, giving the equation

$\mu - 2.25 = 1.656\sigma$ .... [2]

Solving [1] and [2] gives $\mu = 5.78$ and $\sigma = 2.13$ (to 3 significant figures).

$P(W > 6.48) = 1 - P(W \leqslant 6.48) = 1 - \Phi\left(\dfrac{6.48 - 5.7824\ldots}{2.1331\ldots}\right) = 1 - \Phi(0.327) = 0.372$

14 $1 - \Phi\left(\dfrac{147 - \mu}{\sigma}\right) = 0.0136$, so $\dfrac{147 - \mu}{\sigma} = \Phi^{-1}(0.9864) = 2.21$, giving the equation

$147 - \mu = 2.21\sigma$ .... [1]

$1 - \Phi\left(\dfrac{\mu - 59}{\sigma}\right) = 0.0038$, so $\dfrac{\mu - 59}{\sigma} = \Phi^{-1}(0.9962) = 2.67$, giving the equation

$\mu - 59 = 2.67\sigma$ .... [2]

Solving [1] and [2] gives $\mu = 107.147\ldots$ and $\sigma = 18.032\ldots$

$P(80.0 \leqslant X < 130.0) = \Phi\left(\dfrac{130 - 107.147\ldots}{18.032\ldots}\right) - \left[1 - \Phi\left(\dfrac{107.147\ldots - 80}{18.032\ldots}\right)\right]$

$\qquad = \Phi(1.267) - [1 - \Phi(1.505)] = 0.831$

EXERCISE 8D

1 Length, $L \sim N(18.5, 0.7)$, so $P(L < 18.85) = \Phi\left(\dfrac{18.85 - 18.5}{\sqrt{0.7}}\right) = \Phi(0.418) = 0.662$

2 a Waiting time, $T \sim N(13, 16)$, so $P(T > 16.5) = 1 - \Phi\left(\dfrac{16.5 - 13}{\sqrt{16}}\right) = 1 - \Phi(0.875) = 0.191$

 b $P(T < 9) = \Phi\left(\dfrac{9 - 13}{\sqrt{16}}\right) = 1 - \Phi(1) = 0.1587$

 $15.87\% \times 468 \approx 74$ patients wait for less than nine minutes.

3 a Mass in grams, $X \sim N(90, 17.7^2)$

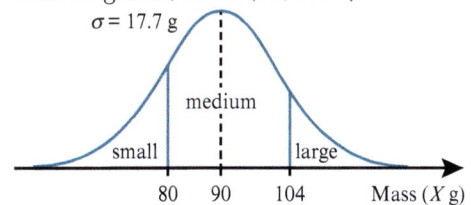

Percentage small $= P(X < 80) = \Phi\left(\dfrac{80 - 90}{17.7}\right) = 1 - \Phi(0.565) = 28.60\%$.

Percentage large $= P(X > 104) = 1 - \Phi\left(\dfrac{104 - 90}{17.7}\right) = 1 - \Phi(0.791) = 21.45\%$.

Percentage medium: $100 - 21.45 - 28.60 = 49.95\%$.

The distribution of the variable can be written as $X \sim N(90, 17.7^2)$ or as $X \sim N(90, 313.29)$.

 b $\Phi\left(\dfrac{104 - 90}{17.7}\right) - \Phi\left(\dfrac{k - 90}{17.7}\right) = 0.75$

$\Phi\left(\dfrac{k - 90}{17.7}\right) = 0.0355$

$\dfrac{90 - k}{17.7} = \Phi^{-1}(1 - 0.0355) = \Phi^{-1}(0.9645) = 1.805 \text{ or } 1.806$

$k = 90 - (17.7 \times 1.805 \text{ or } 1.806) = 58.0 \text{ or } 58.1$

4 Height in metres, $H \sim N(\mu, 3.6^2)$ and $P(H < 10) = 0.75$.

$\Phi\left(\dfrac{10 - \mu}{3.6}\right) = 0.75$, so $\dfrac{10 - \mu}{3.6} = \Phi^{-1}(0.75) = 0.674$, giving $\mu = 7.57$

5 a Mass in kg, $M \sim N(5.73, 2.56)$, so $P(M < 6.0) = \Phi\left(\dfrac{6.0 - 5.73}{\sqrt{2.56}}\right) = \Phi(0.169) = 0.567$

 b $P(M > 3.9) = 1 - \Phi\left(\dfrac{3.9 - 5.73}{\sqrt{2.56}}\right) = \Phi(1.144) = 0.874$

 c $P(7.0 < M < 8.0) = \Phi\left(\dfrac{8.0 - 5.73}{\sqrt{2.56}}\right) - \Phi\left(\dfrac{7.0 - 5.73}{\sqrt{2.56}}\right) = \Phi(1.419) - \Phi(0.794) = 0.136$

6 a Distance in metres, $D \sim N(199, 3700)$, and $P(D < b) = 0.75$.

$\Phi\left(\dfrac{b - 199}{\sqrt{3700}}\right) = 0.75$, so $\dfrac{b - 199}{\sqrt{3700}} = \Phi^{-1}(0.75) = 0.674$

$b = 199 + 0.674 \times \sqrt{3700} = 240$

b The upper quartile is $b = 240$ and the distribution is symmetrical about the mean, so $IQR = 2 \times (240 - 199) = 82.0\,m$.

> Recall that $P(X < Q_1) = 0.25$; $P(X < Q_2) = 0.5$ and $P(X < Q_3) = 0.75$.

7 Daily % change, $X \sim N(0, 0.51^2)$, so

$P(X < -1) = \Phi\left(\dfrac{-1 - 0}{0.51}\right) = 1 - \Phi(1.961) = 0.0249$

They should expect this to occur on $0.0249 \times 365 \approx 9$ days.

8 $P(187 \leqslant w < 213) = 2\Phi(1) - 1 = 0.6826$

Let the number of apples in the sample be n then $0.6826n = 3413$ gives $n = 5000$.

> 187 and 213 are both σ units (one standard deviation) from the mean of 200.

9 Age in years, $A \sim N(15.2, \sigma^2)$ and $P(A < 13.5) = 0.305$

$\Phi\left(\dfrac{13.5 - 15.2}{\sigma}\right) = 0.305$, so

$\Phi\left(\dfrac{15.2 - 13.5}{\sigma}\right) = 1 - 0.305 = 0.695$

$\dfrac{1.7}{\sigma} = \Phi^{-1}(0.695) = 0.51$, giving $\sigma = 3.33$

10 Speed in kmh^{-1}, $S \sim N(\mu, 20^2)$ and $P(S > 100) = 0.33$

$\Phi\left(\dfrac{100 - \mu}{20}\right) = 1 - 0.33 = 0.67$, so

$\dfrac{100 - \mu}{20} = \Phi^{-1}(0.67) = 0.44$, giving $\mu = 91.2$

Percentage under $80\,kmh^{-1}$ is

$P(S < 80) = \Phi\left(\dfrac{80 - 91.2}{20}\right) = 1 - \Phi(0.56) = 28.8\%$

11 Mass in grams, $M \sim N(210, \sigma^2)$ and $P(M < 200) = 0.005$

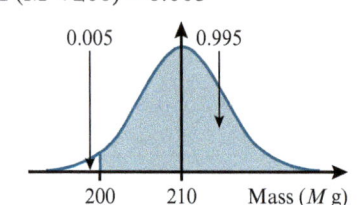

$\Phi\left(\dfrac{200 - 210}{\sigma}\right) = 0.005$, so $\Phi\left(\dfrac{10}{\sigma}\right) = 0.995$

$\dfrac{10}{\sigma} = \Phi^{-1}(0.995) = 2.576$ gives $\sigma = 3.88$

12 a Time in minutes, $T \sim N(12.8, \sigma^2)$ and

$P(T > 15) = \dfrac{42}{365} \approx 0.115$

$\Phi\left(\dfrac{15 - 12.8}{\sigma}\right) = 1 - 0.115 = 0.885$, so

$\dfrac{2.2}{\sigma} = \Phi^{-1}(0.885) = 1.2$ gives $\sigma = 1.83$

b $P(T < 10) = \Phi\left(\dfrac{10 - 12.8}{1.83333}\right) = 1 - \Phi(1.527)$

$= 0.0635$

Colleen is expected to do this on $365 \times 0.0635 \approx 23$ days

13 a Time taken in minutes, $T \sim N(\mu, 7.42^2)$ and $P(T > 20) = 0.75$. $\Phi\left(\dfrac{\mu - 20}{7.42}\right) = 0.75$,

so $\dfrac{\mu - 20}{7.42} = \Phi^{-1}(0.75) = 0.674$

$\mu = 20 + 0.674 \times 7.42 = 25.0$

b $Q_1 = 20$ and $Q_2 = 25$ so, by symmetry, $Q_3 = 30$.
$P(T \geq 30) = 0.25$, so $0.25n = 250$ gives $n = 1000$.

14 a $\Phi\left(\dfrac{\mu + \sigma - \mu}{\sigma}\right) - \Phi\left(\dfrac{\mu - \sigma - \mu}{\sigma}\right)$
$= \Phi(1) - [1 - \Phi(1)] = 2\Phi(1) - 1 = 0.683$

b $1 - \left[\Phi\left(\dfrac{\mu + 2\sigma - \mu}{\sigma}\right) - \Phi\left(\dfrac{\mu - 2\sigma - \mu}{\sigma}\right)\right]$
$= 1 - \{\Phi(2) - [1 - \Phi(2)]\} = 2 - 2\Phi(2) = 0.0456$

c $\Phi\left(\dfrac{7.5 - \mu}{\sigma}\right) = 0.75$, so
$\dfrac{7.5 - \mu}{\sigma} = \Phi^{-1}(0.75) = 0.674$, giving the equation $7.5 - \mu = 0.674\sigma$ .... [1]

$\Phi\left(\dfrac{8.5 - \mu}{\sigma}\right) = 0.9$, so $\dfrac{8.5 - \mu}{\sigma} = \Phi^{-1}(0.9)$
$= 1.282$, giving the equation
$8.5 - \mu = 1.282\sigma$ .... [2]

Solving [1] and [2] gives $\sigma = 1.64$ and $\mu = 6.39$.

15 a Time taken, $T \sim N(9, 5.91)$, so
$P(T \geq 5) = \Phi\left(\dfrac{9 - 5}{\sqrt{5.91}}\right) = \Phi(1.645) = 0.950$

b Let the number of times that the document fails to open in under exactly five seconds be X, then $X \sim B(n, 0.05)$ and $P(X \geq 1) > 0.5$.
$P(X \geq 1) = 1 - P(X = 0) = 1 - \binom{n}{0} \times 0.05^0 \times 0.95^n$
$= 1 - 0.95^n$
Now $1 - 0.95^n > 0.5$
$0.95^n < 0.5$
$n \log 0.95 < \log 0.5$
$n > \dfrac{\log 0.5}{\log 0.95} = 13.513...,$
so the least n is 14.

16 a Mass in grams, $M \sim N(400, 61^2)$, so
$P(M < 425) = \Phi\left(\dfrac{425 - 400}{61}\right) = \Phi(0.410)$
$= 0.659$

b Let the number of pies with masses less than 425 g be X, then $X \sim B(4, 0.6591)$.
$P(X = 4) = \binom{4}{4} \times 0.6591^4 \times 0.3409^0 = 0.189$

c Let the number of pies with masses less than 425 g be Y, then $Y \sim B(10, 0.6591)$.
$P(Y = 7) = \binom{10}{7} \times 0.6591^7 \times 0.3409^3 = 0.257$

17 a Height in metres, $H \sim N(1.74, 0.123^2)$.
$P(1.71 < H < 1.80) = \Phi\left(\dfrac{1.80 - 1.74}{0.123}\right) - \Phi\left(\dfrac{1.71 - 1.74}{0.123}\right) = \Phi(0.488) - [1 - \Phi(0.244)]$
$= 0.284$

b Let the number of females between 1.71 and 1.80 m be X, then $X \sim B(3, 0.2836)$.
$P(X = 3) = \binom{3}{3} \times 0.2836^3 \times 0.7164^0 = 0.0228$

c Let the number of females between 1.71 and 1.80 m be Y, then $Y \sim B(50, 0.2836)$.
$P(Y = 15) = \binom{50}{15} \times 0.2836^{15} \times 0.7164^{35} = 0.118$

EXERCISE 8E

1 a $np = 20 \times 0.6 = 12$ and $nq = 20 \times 0.4 = 8$. It can be well approximated.
$\mu = np = 12$ and $\sigma^2 = npq = 4.8$
b $np = 30 \times 0.95 = 28.5$ and $nq = 30 \times 0.05 = 1.5$. It cannot be well approximated because $nq = 1.5 < 5$.
c $np = 40 \times 0.13 = 5.2$ and $nq = 40 \times 0.87 = 34.8$. It can be well approximated.
$\mu = np = 5.2$ and $\sigma^2 = npq = 4.524$
d $np = 50 \times 0.06 = 3$ and $nq = 50 \times 0.94 = 47$. It cannot be well approximated because $np = 3 < 5$.

2 a $n \times 0.024 > 5$, so $n > 208\frac{1}{3}$ and the least n is 209.

b $n \times 0.15 > 5$, so $n > 33\frac{1}{3}$ and the least n is 34.

c $n \times (1 - 0.52) > 5$, so $n > 10\frac{5}{12}$ and the least n is 11.

d $n \times (1 - 0.7) > 5$, so $n > 16\frac{2}{3}$ and the least n is 17.

3 $q = \dfrac{npq}{np} = \dfrac{10.5}{14} = 0.75$ and $p = 0.25$

$n = \dfrac{np}{p} = \dfrac{14}{0.25} = 56$, so the distribution is

B(56, 0.25).

Note: CC stands for *continuity correction*.

4 $np = 70$, $npq = 21$, so approximate B(100, 0.7) by N(70, 21) with a CC at 74.5.

$P(X < 75) \approx \Phi\left(\dfrac{74.5 - 70}{\sqrt{21}}\right) = \Phi(0.982) = 0.837$

> For $P(X < a)$ and $P(X \geq a)$, the correct CC is at $a - 0.5$.
> For $P(X > a)$ and $P(X \leq a)$, the correct CC is at $a + 0.5$.

5 $np = 30$, $npq = 12$, so approximate B(50, 0.6) by N(30, 12) with a CC at 26.5.

$P(Y > 26) \approx 1 - \Phi\left(\dfrac{26.5 - 30}{\sqrt{12}}\right) = \Phi(1.010)$
$= 0.844$

6 a $p = \dfrac{100}{160} = 0.625$, so

Var(H) $= 160 \times 0.625 \times (1 - 0.625) = 37.5$

b Approximate $H \sim$ B(160, 0.625) by N(100, 37.5) with a CC at 110.5.

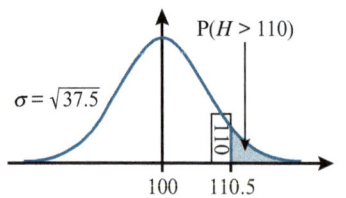

$P(H > 110) \approx 1 - \Phi\left(\dfrac{110.5 - 100}{\sqrt{37.5}}\right)$
$= 1 - \Phi(1.715) = 0.0432$

> The variable H is discrete. To find $P(H > 110)$, we use a normal approximation with a continuity correction rather than calculating $P(H = 111) + P(H = 112) + \ldots + P(H = 159) + P(H = 160)$.

7 a Var(C) $= 40 \times 0.25 \times 0.75 = 7.5$.

b Approximate $C \sim$ B(40, 0.25) by N(10, 7.5) with a CC at 8.5.

$P(C \leq 8) \approx \Phi\left(\dfrac{8.5 - 10}{\sqrt{7.5}}\right) = 1 - \Phi(0.548) = 0.292$

The approximation is justified because $np = 10$ and $nq = 30$ are both greater than 5.

8 a Expected number in full-time employment is $np = 80 \times 0.55 = 44$.

b SD $= \sqrt{npq} = \sqrt{80 \times 0.55 \times 0.45} = 4.45$

c Approximate B(80, 0.55) by N(44, 19.8) with a CC at 39.5.

$P(X < 40) \approx \Phi\left(\dfrac{39.5 - 44}{\sqrt{19.8}}\right) = 1 - \Phi(1.011)$
$= 0.156$

Chapter 8: The normal distribution

9 a i Using $X \sim B(25, 0.8)$, $P(X = 21) = \binom{25}{21} \times 0.8^{21} \times 0.2^4 = 0.187$

ii Using $Y \sim B(25, 0.2)$, $P(Y = 10) = \binom{25}{10} \times 0.2^{10} \times 0.8^{15} = 0.0118$

b Let the number of rubber washers in a retail pack be R.
$E(R) = 2000 \times 0.8 = 1600$; $Var(R) = 2000 \times 0.8 \times 0.2 = 320$

c Approximate $R \sim B(2000, 0.8)$ by $N(1600, 320)$ with a CC at 1620.5.

$P(R \leq 1620) \approx \Phi\left(\dfrac{1620.5 - 1600}{\sqrt{320}}\right) = \Phi(1.146) = 0.874$

10 a i Using $X \sim B(20, 0.63)$, $P(X = 15) = \binom{20}{15} \times 0.63^{15} \times 0.37^5 = 0.105$

ii Using $Y \sim B(20, 0.37)$, $P(Y = 9) = \binom{20}{9} \times 0.37^9 \times 0.63^{11} = 0.135$

b Let the number of homes with an internet connection be H, then $H \sim B(600, 0.63)$, which we approximate by $N(378, 139.86)$ with a CC at 390.5.

$P(H > 390) \approx 1 - \Phi\left(\dfrac{390.5 - 378}{\sqrt{139.86}}\right) = 1 - \Phi(1.057) = 0.145$

11 Let the number of people who watch more than two hours of TV per day be X, then $X \sim B(300, 0.17)$, which we approximate by $N(51, 42.33)$ with a CC at 59.5.

$P(X \geq 60) \approx 1 - \Phi\left(\dfrac{59.5 - 51}{\sqrt{42.33}}\right) = 1 - \Phi(1.306) = 0.0958$

12 a Using $X \sim B(120, 0.41)$, $P(X = 50) = \binom{120}{50} \times 0.41^{50} \times 0.59^{70} = 0.0729$.

b $n = 120$ and $p = 0.41 + 0.23 = 0.64$

We approximate $B(120, 0.64)$ by $N(76.8, 27.648)$ with CCs at 70.5 and 89.5.

$\Phi\left(\dfrac{89.5 - 76.8}{\sqrt{27.648}}\right) - \Phi\left(\dfrac{70.5 - 76.8}{\sqrt{27.648}}\right) = \Phi(2.415) - [1 - \Phi(1.198)] = 0.877$

13 a Let the number of damaged tiles be X, then $X \sim B(38400, 0.0015625)$, which we approximate by $N(60, 59.90625)$ with a CC at 65.5.

$P(X > 65) \approx 1 - \Phi\left(\dfrac{65.5 - 60}{\sqrt{59.90625}}\right) = 1 - \Phi(0.711) = 0.239.$

b Let the number of loads with more than 65 damaged tiles be Y, then $Y \sim B(5, 0.2386)$.

$P(Y = 3) = \binom{5}{3} \times 0.2386^3 \times 0.7614^2 = 0.0787$

14 a Let the number of defective memory sticks be D, then $D \sim B(400, 0.02)$, which we approximate by $N(8, 7.84)$ with CCs at 4.5 and 11.5.

$$P(5 \leq D \leq 11) \approx \Phi\left(\frac{11.5 - 8}{\sqrt{7.84}}\right) - \Phi\left(\frac{4.5 - 8}{\sqrt{7.84}}\right) = \Phi(1.25) - [1 - \Phi(1.25)] = 2\Phi(1.25) - 1 = 0.789$$

b $P(\text{fewer than 12 defective in each sample}) \approx \Phi\left(\frac{11.5 - 8}{\sqrt{7.84}}\right) = \Phi(1.25) = 0.8944$

Let the number of samples with fewer than 12 defective memory sticks be Y, then $Y \sim B(10, 0.8944)$. We cannot approximate this by a normal distribution because $nq = 1.056 < 5$.

$P(Y > 7) = P(Y = 8) + P(Y = 9) + P(Y = 10)$

$= \binom{10}{8} \times 0.8944^8 \times 0.1056^2 + \binom{10}{9} \times 0.8944^9 \times 0.1056^1 + \binom{10}{10} \times 0.8944^{10} \times 0.1056^0$

$= 0.920$

15 Let the number of people that approve be X, then $X \sim B(120, 0.57)$, which we approximate by $N(68.4, 29.412)$ with CCs at 59.5 and 75.5.

$$P(X > 75 \mid X \geq 60) = \frac{P(X \geq 60 \text{ and } X > 75)}{P(X \geq 60)} = \frac{P(X > 75)}{P(X \geq 60)}$$

The required probability expresses the small area (indicated at the right of the following diagram) as a fraction of the larger area at the left.

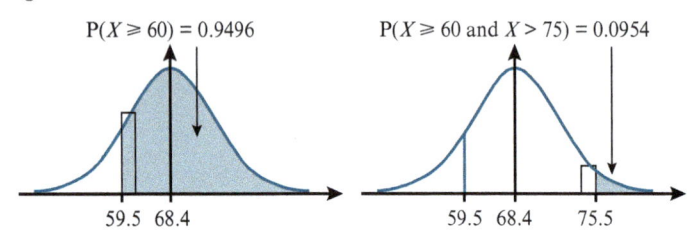

$P(X \geq 60) = 0.9496 \qquad P(X \geq 60 \text{ and } X > 75) = 0.0954$

59.5 68.4 $\qquad\qquad$ 59.5 68.4 75.5

$$\therefore P(X > 75 \mid X \geq 60) \approx \frac{1 - \Phi\left(\frac{75.5 - 68.4}{\sqrt{29.412}}\right)}{1 - \Phi\left(\frac{59.5 - 68.4}{\sqrt{29.412}}\right)} = \frac{1 - \Phi(1.309)}{\Phi(1.641)} = \frac{0.0954}{0.9496} = 0.100$$

16 Let the number of heads be X, then $X \sim B(400, 0.5)$, which we approximate by $N(200, 10^2)$ with CCs at 205.5 and 214.5.

$$P(X < 215 \mid X > 205) = \frac{P(205 < X < 215)}{P(X > 205)} \approx \frac{\Phi\left(\frac{214.5 - 200}{10}\right) - \Phi\left(\frac{205.5 - 200}{10}\right)}{1 - \Phi\left(\frac{205.5 - 200}{10}\right)}$$

$$= \frac{[\Phi(1.45) - \Phi(0.55)]}{1 - \Phi(0.55)} = \frac{0.2177}{0.2912} = 0.748$$

17 Let the number of 6s rolled be S, then $S \sim B\left(450, \frac{1}{6}\right)$, which we approximate by $N(75, 62.5)$ with CCs at 79.5 and 69.5.

$$P(S \geq 70 \mid S < 80) = \frac{P(70 \leq S < 80)}{P(S < 80)} \approx \frac{\Phi\left(\frac{79.5 - 75}{\sqrt{62.5}}\right) - \Phi\left(\frac{69.5 - 75}{\sqrt{62.5}}\right)}{\Phi\left(\frac{79.5 - 75}{\sqrt{62.5}}\right)}$$

$$= \frac{\Phi(0.569) + \Phi(0.696) - 1}{\Phi(0.569)} = \frac{0.4722}{0.7154} = 0.660$$